Papers in philosophical logic

This is part of a three-volume collection of most of David Lewis's papers in philosophy, except for those which previously appeared in his *Philosophical Papers* (Oxford University Press, 1983 and 1986). They are now offered in a readily accessible form.

This first volume is devoted to Lewis's work in philosophical logic. Topics covered include the formal semantics of natural languages; model-theoretic investigations of intensional logic; contradiction and relevance; the distinction between analog and digital representation; attempts to draw anti-mechanistic conclusions from Gödel's theorem; Carnap's *Aufbau*; mereology and its relationship to set theory.

The purpose of this collection, and the two volumes to follow, is to disseminate more widely the work of an eminent and influential contemporary philosopher. The volume will serve as a useful work of reference for teachers and students of philosophy.

T0350166

CAMBRIDGE STUDIES IN PHILOSOPHY

General editor ERNEST SOSA (Brown University)

Advisory editors:
JONATHAN DANCY (University of Keele)
JOHN HALDANE (University of St. Andrews)
GILBERT HARMAN (Princeton University)
FRANK JACKSON (Australian National University)
WILLIAM G. LYCAN (University of North Carolina at Chapel Hill)
SYDNEY SHOEMAKER (Cornell University)
JUDITH J. THOMSON (Massachusetts Institute of Technology)

RECENT TITLES

JAEGWON KIM Supervenience and mind
WARREN QUINN Morality and action
MICHAEL JUBIEN Ontology, morality, and the fallacy of reference
JOHN W. CARROLL Laws of nature
HENRY S. RICHARDSON Practical reasoning about final ends
JOSHUA HOFFMAN and GARY S. ROSENCRANTZ Substance among other categories
M. J. CRESSWELL Language in the world
NOAH LEMOS Intrinsic value
PAUL HELM Belief policies
LYNNE RUDDER BAKER Explaining attitudes
ROBERT A. WILSON Cartesian psychology and physical minds
BARRY MAUND Colours
MICHAEL DEVITT Coming to our senses
MICHAEL ZIMMERMAN The concept of moral obligation
MICHAEL STOCKER with ELIZABETH HEGEMAN Valuing emotions
SYDNEY SHOEMAKER The first-person perspective and other essays
NORTON NELKIN Consciousness and the origins of thought
MARK LANCE and JOHN O'LEARY HAWTHORNE The grammar of meaning
D. M. ARMSTRONG A world of states of affairs
PIERRE JACOB What minds can do
ANDRE GALLOIS The world without the mind within
FRED FELDMAN Utilitarianism, hedonism, and desert
LAURENCE BONJOUR In defense of pure reason

Papers in
philosophical logic

DAVID LEWIS

Princeton University

CAMBRIDGE
UNIVERSITY PRESS

CAMBRIDGE UNIVERSITY PRESS
Cambridge, New York, Melbourne, Madrid, Cape Town, Singapore,
São Paulo, Delhi, Dubai, Tokyo, Mexico City

Cambridge University Press
The Edinburgh Building, Cambridge CB2 8RU, UK

Published in the United States of America by Cambridge University Press, New York

www.cambridge.org
Information on this title: www.cambridge.org/9780521582476

First published 1998

A catalogue record for this publication is available from the British Library

Library of Congress Cataloguing in Publication data
Lewis, David K., 1941–
Papers in philosophical logic/David Lewis.
p.   cm. – (Cambridge studies in philosophy)
Includes bibliographical references and index.
ISBN 0-521-58247-4 (hardback). – ISBN 0-521-58788-3 (pbk.)
1. Logic, Symbolic and mathematical.   I. Title. II. Series.
BC135.L44   1998   97-6656
160–dc21   CIP

ISBN 978-0-521-58247-6 Hardback
ISBN 978-0-521-58788-4 Paperback

Contents

v

For the philosophers, past and present
of Wellington and Uppsala

Introduction

This collection reprints almost all my previously published papers in philosophical logic, except for those that were previously reprinted in another collection, *Philosophical Papers*.[1] Still omitted are (1) two papers in deontic logic; (2) a paper on counterfactual logic,[2] which is superseded by proofs later published in my book *Counterfactuals;*[3] and (3) two papers on immodest inductive methods,[4] the first of them completely mistaken and the second one concerned with loose ends left over when the first was rebutted. I have taken the opportunity to correct typographical and editorial errors. But I have left the philosophical content as it originally was, rather than trying to rewrite the papers as I would write them today. A very few afterthoughts have been noted in new footnotes.

The first four papers have to do with the project of transplanting the methods of formal semantics from artificial formalized languages to natural languages, or to formalized approximations thereof. 'Index, Context, and Content' is a meta-theoretical paper, arguing that some supposedly competing formulations are no more than

---

1 David Lewis, *Philosophical Papers,* volumes I and II (Oxford University Press, 1983 and 1986).
2 'Completeness and Decidability of Three Logics of Counterfactual Conditionals', *Theoria* 37 (1971), pp. 74–85.
3 David Lewis, *Counterfactuals* (Blackwell, 1973).
4 'Immodest Inductive Methods', *Philosophy of Science* 38 (1971), pp. 54–63; 'Spielman and Lewis on Inductive Immodesty', *Philosophy of Science* 41 (1974), pp. 84–85.

1

notational variants. The other three study particular constructions. 'Adverbs of Quantification' examines a neglected part of the quantificational apparatus of ordinary English: sentential adverbs like 'always', sometimes, 'often', *et al.* (I first became interested in these adverbs as a means to making my own technical writing more concise. Only afterward did it occur to me that they afforded a treatment of a class of problematic sentences in which it seemed that there was no satisfactory way to assign relative scopes to quantifier phrases.) ' "Whether" Report' applies the technique of double indexing to treat 'whether'-clauses as sentences that express whichever is the true one of several listed alternatives. 'Probabilities of Conditionals and Conditional Probabilities II' continues my attack (begun in a paper which is reprinted in *Philosophical Papers* and hence not included here) upon the idea that the English conditional connective has a semantic analysis that must be explained in terms of probability values rather than truth values.

The next two papers have to do with model-theoretic investigations of systems of intensional logic. Such investigations usually culminate in completeness results: theorems saying that the sentences provable by means of certain rules and axioms are exactly those sentences that are validated by a certain sort of models. When we have such theorems, we have a complete syntactic, computational test for the presence (though sometimes not also for the absence) of the semantic property of validity. 'Intensional Logics Without Iterative Axioms' gives a general method that produces completeness results wholesale for a certain class of logics, including the most basic systems of modal, deontic, and conditional logic. 'Ordering Semantics and Premise Semantics for Counterfactuals' investigates the relation between two seemingly different conceptions of semantic validity in the logic of counterfactual conditionals.

The next four papers deal with matters of contradiction and relevance. In standard logic, a contradiction is said to imply everything. Thus there are valid arguments in which the premises seem irrelevant to the conclusion: for instance, when a contradiction about the weather implies any conclusion you like about the economy. Most logicians, and I for one, think this situation unproblematic; but some think it very objectionable, and so there is a flourishing industry of building systems of 'relevant logic' in which this supposedly objectionable thing will not happen. In 1980 I visited Canberra, a center for the development and advocacy of relevant logic. I was not at all convinced by the relevantists' case against standard logic; but I agreed that it made sense to want a discriminating notion of what was and wasn't true according to a contradictory corpus of propositions, and I also agreed that it was worthwhile to analyze relationships of relevance. I doubted, however, that these two projects had much to do with one another. In 'Logic for Equivocators' I argue that whether or not we agree with the original motivation for relevant logic, it has produced a technology that serves a different useful purpose: it can protect us from falling victim to fallacies of ambiguity even when it is beyond our power to disambiguate our language fully. In 'Relevant Implication' I argue that we can build a well-motivated formal analysis of relevance, but that this analysis will not deliver relevant logic. In 'Statements Partly About Observation' I argue that this same analysis will deliver a solution to the positivists' old problem of delineating empirical statements from others, though not a solution that will support the positivists' idea that only empirical statements are meaningful. Finally, in 'Ayer's First Empiricist Criterion of Meaning: Why Does it Fail?' I call attention to a minor but oft-repeated mistake about what went wrong with one early attempt at delineating the empirical statements.

The next paper, 'Analog and Digital', addresses a problem set by Nelson Goodman in my student days: what is the difference between analog and digital representation? Goodman was right to fault some familiar answers: once we appreciate just how wide a range of physical magnitudes we can define, for instance, we will not deny that digital computers represent quantities by means of physical magnitudes. But Goodman's own answer[5] seemed to me to go wrong as well.

The next two papers, 'Lucas Against Mechanism' and its sequel, reply to different versions of J.R. Lucas's allegation that Gödel's theorem shows that our mathematical powers exceed those of any machine. Lucas is right about what machines cannot do, but fails to make his case that we can do better.

5 In *Language of Art* (Bobbs-Merrill, 1968), p. 160.

The four remaining papers address questions arising in the construction of ambitious formalized philosophical systems, and again they reflect the influence of Nelson Goodman. Carnap's *Aufbau*[6] is perhaps the most ambitious of all these systems, but it failed on its own terms. In 'Policing the *Aufbau*' I show how its performance might be improved – though not to the point where success is guaranteed.

The *Aufbau* was built upon a framework of set theory, in much the way that modern mathematics normally is. But, mainly because of the set-theoretical paradoxes, it would be nice to replace set theory by mereology, the formal theory of the relation of part and whole. This is done, for instance, in Goodman's *Structure of Appearance*.[7] Still nicer would be to rebuild set theory itself by mereological means. In 'Finitude and Infinitude in the Atomic Calculus of Individuals', Hodges and I show an obstacle to this ambition: first-order mereology cannot express the difference between a finite and an infinite world. 'Nominalistic Set Theory' shows how to build approximate simulations of set theory within first-order mereology if we have a way to talk about the adjacency or non-adjacency of atoms. But if we give ourselves *second*-order mereology, 'megethology', we can do much better: we can exactly recapture standard set theory, and hence standard mathematics. In 'Mathematics is Megethology'[8] I show how this can be done without falling into known paradoxes, and without allowing our framework to foist upon us any ontological commiments over and above those of the mathematics we set out to recapture.

David Lewis
Princeton, November 1996

6 Rudolf Carnap, *Der Logische Aufbau der Welt* (Weltkreis-Verlag, 1928).
7 Nelson Goodman, *The Structure of Appearance* (Harvard University Press, 1951).
8 Partly an abridgement of material from David Lewis, *Parts of Classes* (Blackwell, 1991).

1

Adverbs of quantification

CAST OF CHARACTERS

The adverbs I wish to consider fall into six groups of near-synonyms, as follows.

(1) Always, invariably, universally, without exception
(2) Sometimes, occasionally, [once]
(3) Never
(4) Usually, mostly, generally, almost always, with few exceptions, [ordinarily], [normally]
(5) Often, frequently, commonly
(6) Seldom, infrequently, rarely, almost never

Bracketed items differ semantically from their list-mates in ways I shall not consider here; omit them if you prefer.

FIRST GUESS: QUANTIFIERS OVER TIMES?

It may seem plausible, especially if we stop with the first word on each list, that these adverbs function as quantifiers over times. That is to say that *always*, for instance, is a modifier that combines with a sentence Φ to make a sentence *Always* Φ that is true iff the modified

First published in Edward L. Keenan (ed.) *Formal Semantics of Natural Language* (Cambridge, Cambridge University Press, 1975). © Cambridge University Press 1975. Reprinted with kind permission from Cambridge University Press.

sentence Φ is true at all times. Likewise, we might guess that *Sometimes* Φ, *Never* Φ, *Usually* Φ, *Often* Φ, and *Seldom* Φ are true, respectively, iff Φ is true at some times, none, most, many, or few. But it is easy to find various reasons why this first guess is too simple.

First, we may note that the times quantified over need not be moments of time. They can be suitable stretches of time instead. For instance,

(7) The fog usually lifts before noon here

means that the sentence modified by *usually* is true on most days, not at most moments. Indeed, what is it for that sentence to be true at a moment?

Second, we may note that the range of quantification is often restricted. For instance,

(8) Caesar seldom awoke before dawn

is not made true by the mere fact that few of all times (past, present, or future) are times when Caesar was even alive, wherefore fewer still are times when he awoke before dawn. Rather it means that few of all the times when Caesar awoke are times before dawn; or perhaps that on few of all the days of his life did he awake before dawn.

Third, we may note that the entities we are quantifying over, unlike times,[1] may be distinct although simultaneous. For instance,

(9) Riders on the Thirteenth Avenue line seldom find seats

may be true even though for 22 hours out of every 24 – all but the two peak hours when 86% of the daily riders show up – there are plenty of seats for all.

SECOND GUESS: QUANTIFIERS OVER EVENTS?

It may seem at this point that our adverbs are quantifiers, suitably restricted, over events; and that times enter the picture only because events occur at times. Thus (7) could mean that most of the daily fog-

---

1 Unlike genuine moments or stretches of time, that is. But we may truly say that Miles the war hero has been wounded 100 times if he has suffered 100 woundings,

liftings occurred before noon; (8) could mean that few of Caesar's awakenings occurred before dawn; and (9) could mean that most rides on the Thirteenth Avenue line are seatless. So far, so good; but further difficulties work both against our first guess and against this alternative.

Sometimes it seems that we quantify not over single events but over enduring states of affairs. For instance,

(10) A man who owns a donkey always beats it now and then

means that every continuing relationship between a man and his donkey is punctuated by beatings; but these continuing relationships, unlike the beatings, are not events in any commonplace sense. Note also that if *always* were a quantifier over times, the sentence would be inconsistent: it would say that the donkey-beatings are incessant and that they only happen now and then. (This sentence poses other problems that we shall consider later.)

We come last to a sweeping objection to both of our first two guesses: the adverbs of quantification may be used in speaking of abstract entities that have no location in time and do not participate in events. For instance,

(11) A quadratic equation never has more than two solutions
(12) A quadratic equation usually has two different solutions

mean, respectively, that no quadratic equation has more than two solutions and that most − more precisely, all but a set of measure zero under the natural measure on the set of triples of coefficients − have two different solutions. These sentences have nothing at all to do with times or events.

Or do they? This imagery comes to mind: someone is contemplating quadratic equations, one after another, drawing at random from all the quadratic equations there are. Each one takes one unit of time. In no unit of time does he contemplate a quadratic equation with more than two solutions. In most units of time he contemplates quadratic equations with two different solutions.

even if he has been wounded at only 99 distinct moments (or stretches) of time because two of his woundings were simultaneous.

For all I know, such imagery may sustain the usage illustrated by (11) and (12), but it offers no hope of a serious analysis. There can be no such contemplator. To be more realistic, call a quadratic equation *simple* iff each of its coefficients could be specified somehow in less than 10,000 pages; then we may be quite sure that the only quadratic equations that are ever contemplated are simple ones. Yet

(13) Quadratic equations are always simple

is false, and in fact they are almost never simple.

THIRD GUESS: QUANTIFIERS OVER CASES

What we can say, safely and with full generality, is that our adverbs of quantification are quantifiers over cases. What holds always, sometimes, never, usually, often, or seldom is what holds in, respectively, all, some, no, most, many, or few cases.

But we have gained safety by saying next to nothing. What is a case? It seems that sometimes we have a case corresponding to each moment or stretch of time, or to each in some restricted class. But sometimes we have a case for each event of some sort; or for each continuing relationship between a man and his donkey; or for each quadratic equation; or − as in the case of this very sentence − for each sentence that contains one of our adverbs of quantification.

UNSELECTIVE QUANTIFIERS

It will help if we attend to our adverbs of quantification as they can appear in a special dialect: the dialect of mathematicians, linguists, philosophers, and lawyers, in which variables are used routinely to overcome the limitations of more colloquial means of pronominalization. Taking m, n, p as variables over natural numbers, and x, y, z as variables over persons, consider:

(14) Always, p divides the product of m and n only if some factor of p divides m and the quotient of p by that factor divides n
(15) Sometimes, p divides the product of m and n although p divides neither m nor n

(16) Sometimes it happens that x sells stolen goods to y, who sells them to z, who sells them back to x

(17) Usually, x reminds me of y if and only if y reminds me of x

Here it seems that if we are quantifying over cases, then we must have a case corresponding to each admissible assignment of values to the variables that occur free in the modified sentence. Thus (14) is true iff every assignment of natural numbers as values of m, n, and p makes the open sentence after *always* true – in other words, iff all triples of natural numbers satisfy that open sentence. Likewise (15) is true iff some triple of numbers satisfies the open sentence after *sometimes*; (16) is true iff some triple of persons satisfies the open sentence after *sometimes*; and (17) is true iff most pairs of persons satisfy the open sentence after *usually*.

The ordinary logicians' quantifiers are selective: $\forall x$ or $\exists x$ binds the variable x and stops there. Any other variables y, z, ... that may occur free in its scope are left free, waiting to be bound by other quantifiers. We have the truth conditions:

(18) $\forall x \Phi$ is true, under any admissible assignment f of values to all variables free in Φ except x, iff for every admissible value of x, Φ is true under the assignment of that value to x together with the assignment f of values to the other variables free in Φ;

(19) $\exists x \Phi$ is true, under any admissible assignment f of values to all variables free in Φ except x, iff for some admissible value of x, Φ is true under the assignment of that value to x together with the assignment f of values to the other variables free in Φ;

and likewise for the quantifiers that select other variables.

It is an entirely routine matter to introduce *unselective quantifiers* $\forall$ and $\exists$ that bind all the variables in their scope indiscriminately. Without selectivity, the truth conditions are much simpler; with no variables left free, we need not relativize the truth of the quantified sentence to an assignment of values to the remaining free variables.

(20) $\forall \Phi$ is true iff Φ is true under every admissible assignment of values to all variables free in Φ;

(21) $\exists\Phi$ is true iff Φ is true under some admissible assignment of values to all variables free in Φ

These unselective quantifiers have not deserved the attention of logicians, partly because they are unproblematic and partly because strings of ordinary, selective quantifiers can do all that they can do, and more besides. They have only the advantage of brevity. Still, brevity *is* an advantage, and it should be no surprise if unselective quantifiers are used in natural language to gain that advantage. That is what I claim; the unselective $\forall$ and $\exists$ can show up as the adverbs *always*, and *sometimes*.[2] Likewise *never, usually, often*, and *seldom* can serve as the unselective analogs of the selective quantifiers *for no x, for most x, for many x*, and *for few x*.[3]

To summarize, what we have in the variable-using dialect is roughly as follows. Our adverbs are quantifiers over cases; a case may be regarded as the 'tuple of its participants; and these participants are values of the variables that occur free in the open sentence modified by the adverb. In other words, we are taking the cases to be the admissible assignments of values to these variables.

But matters are not quite that simple. In the first place, we may wish to quantify past our adverbs, as in

(22) There is a number q such that, without exception, the product of m and n divides q only if m and n both divide q

So our adverbs of quantification are not entirely unselective: they can bind indefinitely many free variables in the modified sentence, but

---

2 It is pleasing to find that Russell often explained the now-standard selective quantifiers by using an unselective adverb of quantification to modify an open sentence. For instance in *Principia* 1, *9, we find the first introduction of quantifiers in the formal development: 'We shall denote "Φx *always*" by the notation $(x).\Phi x$ . . . We shall denote "Φx *sometimes*" by the notation $(\exists x).\Phi x$.'

3 It is customary to work with assignments of values to all variables in the language; the part of the assignment that assigns values to unemployed variables is idle but harmless. But for us this otherwise convenient practice would be more bother than it is worth. In dealing with *usually, often*, and *seldom* we must consider the fraction of value-assignments that satisfy the modified sentence. Given infinitely many variables, these fractions will be ∞/∞ (unless they are 0 or 1). We would need to factor out differences involving only the idle parts of assignments.

some variables – the ones used to quantify past the adverbs – remain unbound. In (22), *m* and *n* are bound by *without exception;* but *q* is immune, and survives to be bound by *there is a number q such that,* a selective quantifier of larger scope.

In the second place, we cannot ignore time altogether in (16)–(17) as we can in the abstract cases (11)–(15); (16)–(17) are not confined to the present moment, but are general over time as well as over 'tuples of persons. So we must treat the modified sentence as if it contained a free time-variable: the truth of the sentence depends on a time-coordinate just as it depends on the values of the person-variables, and we must take the cases to include this time coordinate as well as a 'tuple of persons. (Indeed, we could go so far as to posit an explicit time-variable in underlying structure, in order to subsume time-dependence under dependence on values of variables.) Our first guess about the adverbs is revived as a special case: if the modified sentence has no free variables, the cases quantified over will include nothing but the time coordinate. As noted before, the appropriate time-coordinates (accompanied by 'tuples or not, as the case may be) could either be moments of time or certain stretches of time, for instance days.

Sometimes we might prefer to treat the modified sentence as if it contained an event-variable (or even posit such a variable in underlying structure) and include an event-coordinate in the cases. The event-coordinate could replace the time-coordinate, since an event determines the time of its occurrence. If so, then our second guess also is revived as a special case: if there are no free variables, the cases might simply be events.

In the third place, not just any 'tuple of values of the free variables, plus perhaps a time- or event-coordinate, will be admissible as one of the cases quantified over. Various restrictions may be in force, either permanently or temporarily. Some standing restrictions involve the choice of variables: it is the custom in mathematics that λ is a variable that can take only limit ordinals as values (at least in a suitable context). I set up semi-permanent restrictions of this kind a few paragraphs ago by writing

(23) Taking *m, n, p* as variables over natural numbers, and *x, y,* and *z* as variables over persons . . .

Other standing restrictions require the participants in a case to be suitably related. If a case is a 'tuple of persons plus a time-coordinate, we may take it generally that the persons must be alive at the time to make the case admissible. Or if a case is a 'tuple of persons plus an event-coordinate, it may be that the persons must take part in the event to make the case admissible. It may also be required that the participants in the 'tuple are all different, so that no two variables receive the same value. (I am not sure whether these restrictions are always in force, but I believe that they often are.)

RESTRICTION BY IF-CLAUSES

There are various ways to restrict the admissible cases temporarily – perhaps only for the duration of a single sentence, or perhaps through several sentences connected by anaphoric chains. If-clauses seem to be the most versatile device for imposing temporary restrictions. Consider:

(24) Always, if x is a man, if y is a donkey, and if x owns y, x beats y now and then

A case is here a triple: a value for x, a value for y, and a time-coordinate (longish stretches seem called for, perhaps years). The admissible cases are those that satisfy the three if-clauses. That is, they are triples of a man, a donkey, and a time such that the man owns the donkey at the time. (Our proposed standing restrictions are redundant. If the man owns the donkey at the time, then both are alive at the time; if the participants are a man and a donkey, they are different.) Then (24) is true iff the modified sentence

(25) x beats y now and then

is true in all admissible cases. Likewise for

(26) Sometimes⎫
(27) Usually ⎬ if x is a man, if y is a donkey, and if x owns y,
(28) Often ⎭ x beats y now and then

12

(29) Never ⎱if x is a man, if y is a donkey, and if x owns y,
(30) Seldom ⎰ does x beat y now and then

The admissible cases are the triples that satisfy the if-clauses, and the sentence is true iff the modified sentence (25) – slightly transformed in the negative cases (29)–(30) – is true in some, most, many, none, or few of the admissible cases.

It may happen that every free variable of the modified sentence is restricted by an if-clause of its own, as in

(31) Usually, if x is a man, if y is a donkey, and if z is a dog, y weighs less than x but more than z

But in general, it is best to think of the if-clauses as restricting whole cases, not particular participants therein. We may have any number of if-clauses – including zero, as in (14)–(17). A free variable of the modified sentence may appear in more than one if-clause. More than one variable may appear in the same if-clause. Or it may be that no variable appears in an if-clause; such if-clauses restrict the admissible cases by restricting their time-coordinates (or perhaps their event-co-ordinates), as in

(32) Often if it is raining my roof leaks

(in which the time-coordinate is all there is to the case) or

(33) Ordinarily if it is raining, if x is driving and sees y walking, and if y is x's friend, x offers y a ride

It makes no difference if we compress several if-clauses into one by means of conjunction or relative clauses. The three if-clauses in (24) or in (26)–(30) could be replaced by any of:

(34) if x is a man, y is a donkey, and x owns y . . .
(35) if x is a man and y is a donkey owned by x . . .
(36) if x is a man who owns y, and y is a donkey . . .
(37) if x and y are a man and his donkey . . .

Such compression is always possible, so we would not have gone far wrong to confine our attention, for simplicity, to the case of restriction by a single if-clause.

We have a three-part construction: the adverb of quantification, the if-clauses (zero or more of them), and the modified sentence. Schematically, for the case of a single if-clause:

(38)
$$\left\{ \begin{array}{c} \text{Always} \\ \text{Sometimes} \\ \cdot \\ \cdot \\ \cdot \end{array} \right\} + \text{if}\,\Psi + \Phi$$

But could we get the same effect by first combining Ψ and Φ into a conditional sentence, and then taking this conditional sentence to be the sentence modified by the adverb? On this suggestion (38) is to be regrouped as

(39)
$$\left\{ \begin{array}{c} \text{Always} \\ \text{Sometimes} \\ \cdot \\ \cdot \\ \cdot \end{array} \right\} + \text{if}\,\Psi, \Phi$$

Sentence (39) is true iff the conditional *If* Ψ, Φ is true in all, some, none, most, many, or few of the admissible cases − that is, of the cases that satisfy any permanent restrictions, disregarding the temporary restrictions imposed by the if-clause. But is there any way to interpret the conditional *If* Ψ, Φ that makes (39) equivalent to (38) for all six groups of our adverbs? No; if the adverb is *always* we get the proper equivalence by interpreting it as the truth-functional conditional $\Psi \supset \Phi$, whereas if the adverb is *sometimes* or *never*, that does not work, and we must instead interpret it as the conjunction $\Phi \,\&\, \Psi$. In the remaining cases, there is no natural interpretation that works. I conclude that the *if* of our restrictive if-clauses should not be regarded as a sentential connective. It has no meaning apart from the adverb it restricts. The *if* in *always if . . . , . . . , sometimes if . . . , . . . ,*

14

and the rest is on a par with the non-connective *and* in *between . . . and . . .* , with the non-connective *or* in *whether . . . or . . .* , or with the non-connective *if* in *the probability that . . . if . . . .* It serves merely to mark an argument-place in a polyadic construction.[4]

STYLISTIC VARIATION

Sentences made with the adverbs of quantification need not have the form we have considered so far: adverb + if-clauses + modified sentence. We will find it convenient, however, to take that form — somewhat arbitrarily — as canonical, and to regard other forms as if they were derived from that canonical form. Then we are done with semantics: the interpretation of a sentence in canonical form carries over to its derivatives.

The constituents of the sentence may be rearranged:

(40) If x and y are a man and a donkey and if x owns y, x usually beats y now and then

(41) If x and y are a man and a donkey, usually x beats y now and then if x owns y

(42) If x and y are a man and a donkey, usually if x owns y, x beats y now and then

(43) Usually x beats y now and then, if x and y are a man and a donkey and x owns y

All of (40)–(43), though clumsy, are intelligible and well formed.

Our canonical restrictive if-clauses may, in suitable contexts, be replaced by when-clauses:

---

4 What is the price of forcing the restriction-marking *if* to be a sentential connective after all? Exorbitant: it can be done if (1) we use a third truth value, (2) we adopt a far-fetched interpretation of the connective *if*, and (3) we impose an additional permanent restriction on the admissible cases. Let *If* Ψ, Φ have the same truth value as Φ if Ψ is true, and let it be third-valued if Ψ is false or third-valued. Let a case be admissible only if it makes the modified sentence either true or false, rather than third-valued. Then (39) is equivalent to (38) for all our adverbs, as desired, at least if we assume that Ψ and Φ themselves are not third-valued in any case. A treatment along similar lines of if-clauses used to restrict ordinary, selective quantifiers may be found in Belnap (1970).

(44) When m and n are positive integers, the power m^n can always be computed by successive multiplications

Indeed, a when-clause may sound right when the corresponding if-clause would be questionable, as in a close relative of (8):

(45) Seldom was it before dawn $\begin{Bmatrix} \text{when} \\ ? \text{ if} \end{Bmatrix}$ Caesar awoke

Or we may have a where-clause or a participle construction, especially if the restrictive clause does not come at the beginning of the sentence:

(46) The power m^n, where m and n are positive integers, can always be computed by successive multiplications
(47) The power m^n (m and n being positive integers) can always be computed by successive multiplications

Always if – or is it *always when?* – may be contracted to *whenever*, a complex unselective quantifier that combines two sentences:

(48) Whenever m and n are positive integers, the power m^n can be computed by successive multiplications
(49) Whenever x is a man, y is a donkey, and x owns y, x beats y now and then
(50) Whenever it rains it pours

Always may simply be omitted:

(51) (Always) When it rains, it pours
(52) (Always) If x is a man, y is a donkey, and x owns y, x beats y now and then
(53) When m and n are positive integers, the power m^n can (always) be computed by successive multiplications

Thus we reconstruct the so-called 'generality interpretation' of free variables: the variables are bound by the omitted *always*.

Our stylistic variations have so far been rather superficial. We turn next to a much more radical transformation of sentence structure – a transformation that can bring us back from the variable-using dialect to everyday language.

Suppose that one of our canonical sentences has a restrictive if-clause of the form

(54) if α is τ . . . ,

where α is a variable and τ is an indefinite singular term formed from a common noun (perhaps plus modifiers) by prefixing the indefinite article or *some*.
Examples:

(55) if x is a donkey . . .
(56) if x is an old, grey donkey . . .
(57) if x is a donkey owned by y . . .
(58) if x is some donkey that y owns . . .
(59) if x is something of y's . . .
(60) if x is someone foolish . . .

(Call τ, when so used, a *restrictive term*.) Then we can delete the if-clause and place the restrictive term τ in apposition to an occurrence of the variable α elsewhere in the sentence. This occurrence of α may be in the modified sentence, or in another if-clause of the form (54), or in an if-clause of some other form. Often, but not always, the first occurrence of α outside the deleted if-clause is favoured. If τ is short, it may go before α; if long, it may be split and go partly before and partly after; and sometimes it may follow α parenthetically. The process of displacing restrictive terms may – but need not – be repeated until no if-clauses of the form (54) are left. For instance:

(61) Sometimes, if x is some man, if y is a donkey, and if x owns y, x beats y now and then
$\Rightarrow$

Sometimes if y is a donkey, and if some man x owns y, x beats y now and then

$\Rightarrow$

Sometimes, if some man x owns a donkey y, x beats y now and then

(62) Often, if x is someone who owns y, and if y is a donkey, x beats y now and then

$\Rightarrow$

Often, if x is someone who owns y, a donkey, x beats y now and then

$\Rightarrow$

Often, someone x who owns y, a donkey, beats y now and then

Instead of just going into apposition with an occurrence of the variable α, the restrictive term τ may replace an occurrence of α altogether. Then all other occurrences of α must be replaced as well, either by pronouns of the appropriate case and gender or by terms *that* ν or *the* ν, where ν is the principal noun in the term τ. For instance,

(63) Always, if y is a donkey and if x is a man who owns y, x beats y now and then

$\Rightarrow$

Always, if x is a man who owns a donkey, x beats it now and then

$\Rightarrow$

Always, a man who owns a donkey beats it now and then

Now it is a small matter to move *always* and thereby derive the sentence (10) that we considered earlier. Sure enough, the canonical sentence with which the derivation (63) began has the proper meaning for (10). It is in this way that we return from the variable-using dialect to an abundance of everyday sentences.

I conclude with some further examples.

(64) Always, if x is someone foolish, if y is some good idea, and if x has y, nobody gives x credit for y

$\Rightarrow$

Always, if y is some good idea, and if someone foolish has y, nobody gives him credit for y

$\Rightarrow$

Always, if someone foolish has some good idea, nobody gives him credit for that idea

(65) Often, if y is a donkey, if x is a man who owns y, and if y kicks x, x beats y

$\Rightarrow$

Often, if y is a donkey, and if y kicks a man who owns y, he beats y

$\Rightarrow$

Often, if a donkey kicks a man who owns it, he beats it

(66) Often, if y is a donkey, if x is a man who owns y, and if y kicks x, x beats y

$\Rightarrow$

Often, if x is a man who owns a donkey, and if it kicks x, x beats it

$\Rightarrow$

Often, if it kicks him, a man who owns a donkey beats it

(67) Usually, if x is a man who owns y and if y is a donkey that kicks x, x beats y

$\Rightarrow$

Usually, if x is a man who owns a donkey that kicks x, x beats it

$\Rightarrow$

Usually, a man who owns a donkey that kicks him beats it

(68) Usually, if x is a man who owns y and if y is a donkey that kicks x, x beats y

$\Rightarrow$

Usually, if y is a donkey that kicks him, a man who owns y beats y

$\Rightarrow$

Usually, a man who owns it beats a donkey that kicks him

19

REFERENCES

Belnap, N. (1970). 'Conditional assertion and restricted quantification', *Noûs*, **4,** 1–12.

Russell, B. and Whitehead, A. N. (1912). *Principia Mathematica,* 1. London: Cambridge University Press.

2

Index, context, and content

1. SYNOPSIS

If a grammar is to do its jobs as part of a systematic restatement of our common knowledge about our practices of linguistic communication, it must assign semantic values that determine which sentences are true in which contexts. If the semantic values of sentences also serve to help determine the semantic values of larger sentences having the given sentence as a constituent, then also the semantic values must determine how the truth of a sentence varies when certain features of context are shifted, one feature at a time.

Two sorts of dependence of truth on features of context are involved: *context-dependence* and *index-dependence*. A *context* is a location – time, place, and possible world – where a sentence is said. It has countless features, determined by the character of the location. An *index* is an *n*-tuple of features of context, but not necessarily features that go together in any possible context. Thus an index might consist of a speaker, a time before his birth, a world where he never lived at all, and so on. Since we are unlikely to think of all the features of

First published in Stig Kanger and Sven Öhman (eds.) *Philosophy and Grammar* (Dordrecht, Reidel, 1980). Copyright © 1980 by D. Reidel Publishing Company. Reprinted with kind permission from Kluwer Academic Publishers.

I am grateful to many friends for valuable discussions of the material in this paper. Special mention is due to Max Cresswell, Gareth Evans, Charles Fillmore, David Kaplan, and Robert Stalnaker. An early version was presented to the Vacation School in Logic at Victoria University of Wellington in August 1976; I thank the New Zealand–United States Educational Foundation for research support on that occasion.

context on which truth sometimes depends, and hence unlikely to construct adequately rich indices, we cannot get by without context-dependence as well as index-dependence. Since indices but not contexts can be shifted one feature at a time, we cannot get by without index-dependence as well as context-dependence. An assignment of semantic values must give us the relation: sentence s is true at context c at index i, where i need not be the index that gives the features of context c. Fortunately, an index used together with a context in this way need not give all relevant features of context; only the shiftable features, which are much fewer.

Two alternative strategies are available. (1) Variable but simple semantic values: a sentence has different semantic values at different contexts, and these semantic values are functions from indices to truth values. (2) Complex but constant semantic values: a sentence has the same semantic value at all contexts, and this value is a function from context-index pairs to truth values. But the strategies are not genuine alternatives. They differ only superficially. Hence attempts to argue for the superiority of one over the other are misguided. Whatever one can do, the other can do, and with almost equal convenience.

2. PHILOSOPHY AND GRAMMAR

We have made it part of the business of philosophy to set down, in an explicit and systematic fashion, the broad outlines of our common knowledge about the practice of language. Part of this restatement of what we all know should run as follows. The foremost thing we do with words is to impart information, and this is how we do it. Suppose (1) that you do not know whether A or B or . . . ; and (2) that I do know; and (3) that I want you to know; and (4) that no extraneous reasons much constrain my choice of words; and (5) that we both know that the conditions (1)–(5) obtain. Then I will be truthful and you will be trusting and thereby you will come to share my knowledge. I will find something to say that depends for its truth on whether A or B or . . . and that I take to be true. I will say it and you will hear it. You, trusting me to be willing and able to tell the truth, will then be in a position to infer whether A or B or. . . .

That was not quite right. Consider the tribe of Liars – the ones in

22

the riddles, the folk we fear to meet at forks in the road. Unlike common liars, the Liars have no wish to mislead. They convey information smoothly to each other; and once we know them for what they are, we too can learn from them which road leads to the city. They are as truthful in their own way as we are in ours. But they are truthful in Liarese and we are truthful in English, and Liarese is a language like English but with all the truth values reversed. The missing part of my story concerns our knowledge that we are not of the tribe of Liars. I should not have spoken simply of my truthfulness and your trust. I should have said: I will be truthful-in-English and you will be trusting-in-English, and that is how you will come to share my knowledge. I will find something to say that depends for its truth-in-English on whether A or B or . . . and that I take to be true-in-English; you will trust me to be willing and able to tell the truth-in-English. Truthfulness-in-Liarese would have done as well (and truthfulness-in-English would not have done) had you been trusting-in-Liarese.

Truth-in-English − what is that? A complete restatement of our common knowledge about the practice of language may not use this phrase without explaining it. We need a chapter which culminates in a specification of the conditions under which someone tells the truth-in-English. I call that chapter a *grammar* for English.

I use the word 'grammar' in a broad sense. Else I could have found little to say about our assigned topic. If it is to end by characterizing truth-in-English, a grammar must cover most of what has been called syntax, much of what has been called semantics, and even part of the miscellany that has been called pragmatics. It must cover the part of pragmatics that might better have been called indexical semantics − pragmatics in the sense of Bar-Hillel [1] and Montague [10]. It need not cover some other parts of pragmatics: conversational appropriateness and implicature, disambiguation, taxonomy of speech acts, or what it is about us that makes some grammars right and others wrong.

I am proposing both a delineation of the subject of grammar and a modest condition of adequacy for grammars. A good grammar is one suited to play a certain role in a systematic restatement of our common knowledge about language. It is the detailed and parochial part − the part that would be different if we were Liars, or if we were

Japanese. It attaches to the rest by way of the concept of truth-in-English (or in some other language), which the grammar supplies and which the rest of the restatement employs.

The subject might be differently delineated, and more stringent conditions of adequacy might be demanded. You might insist that a good grammar should be suited to fit into a psycholinguistic theory that goes beyond our common knowledge and explains the inner mechanisms that make our practice possible. There is nothing wrong in principle with this ambitious goal, but I doubt that it is worthwhile to pursue it in our present state of knowledge. Be that as it may, it is certainly not a goal I dare pursue.

3. CONTEXT-DEPENDENCE

Any adequate grammar must tell us that truth-in-English depends not only on what words are said and on the facts, but also on features of the situation in which the words are said. The dependence is surprisingly multifarious. If the words are 'Now I am hungry.' then some facts about who is hungry when matter, but also it matters when the speech occurs and who is speaking. If the words are 'France is hexagonal.' of course the shape of France matters, but so do the aspects of previous discourse that raise or lower the standards of precision. Truth-in-English has been achieved if the last thing said before was 'Italy is sort of boot-shaped.' but not if the last thing said before was 'Shapes in geometry are ever so much simpler than shapes in geography'. If the words are 'That one costs too much.' of course the prices of certain things matter, and it matters which things are traversed by the line projected from the speaker's pointing finger, but also the relations of comparative salience among these things matter. These relations in turn depend on various aspects of the situation, especially the previous discourse. If the words are 'Fred came floating up through the hatch of the spaceship and turned left.', then it matters what point of reference and what orientation we have established. Beware: these are established in a complicated way. (See Fillmore [3].) They need not be the location and orientation of the speaker, or of the audience, or of Fred, either now or at the time under discussion.

When truth-in-English depends on matters of fact, that is called

24

contingency. When it depends on features of context, that is called *indexicality.* But need we distinguish? Some contingent facts are facts about context, but are there any that are not? Every context is located not only in physical space but also in logical space. It is at some particular possible world – our world if it is an actual context, another world if it is a merely possible context. (As you see, I presuppose a metaphysics of modal realism. It's not that I think this ontological commitment is indispensable to proper philosophy of language – in philosophy there are usually many ways to skin a cat. Rather, I reject the popular presumption that modal realism stands in need of justification.) It is a feature of any context, actual or otherwise, that its world is one where matters of contingent fact are a certain way. Just as truth-in-English may depend on the time of the context, or the speaker, or the standards of precision, or the salience relations, so likewise may it depend on the world of the context. Contingency is a kind of indexicality.

4. SEMANTIC VALUES

A concise grammar for a big language – for instance, a finite grammar for an infinite language like ours – had better work on the compositional principle. Most linguistic expressions must be built up stepwise, by concatenation or in some more complicated way, from a stock of basic expressions.

(Alternatively, structures that are not linguistic expressions may be built up stepwise, and some of these may be transformed into linguistic expressions. For evidence that these approaches differ only superficially, see Cooper and Parsons [4].)

To go beyond syntax, a compositional grammar must associate with each expression an entity that I shall call its *semantic value.* (In case of ambiguity, more than one must be assigned.) These play a twofold role. First, the semantic values of some expressions, the *sentences,* must enter somehow into determining whether truth-in-English would be achieved if the expression were uttered in a given context. Second, the semantic value of any expression is to be determined by the semantic values of the (immediate) constituents from which it is built, together with the way it is built from them.

To the extent that sentences are built up, directly or indirectly, from sentences, the semantic values of sentences have both jobs to do. The semantic values of non-sentences have only one job: to do their bit toward determining the semantic values of the sentences.

Semantic values may be anything, so long as their jobs get done. Different compositional grammars may assign different sorts of semantic values, yet succeed equally well in telling us the conditions of truth-in-English and therefore serve equally well as chapters in the systematic restatement of our common knowledge about language. Likewise, different but equally adequate grammars might parse sentences into different constituents, combined according to different rules.

More ambitious goals presumably would mean tighter constraints. Maybe a grammar that assigns one sort of semantic value could fit better into future psycholinguistics than one that assigns another sort. Thereof I shall not speculate.

Another source of obscure and unwanted constraints is our traditional semantic vocabulary. We have too many words for semantic values, and for the relation of having a semantic value:

apply to	express	represent
Bedeutung	extension	satisfy
character	fall under	sense
comply with	intension	signify
comprehension	interpretation	*Sinn*
concept	meaning	stand for
connotation	name	statement
denote	nominatum	symbolize
designate	refer	true of

for a start. Not just any of these words can be used for just any sort of assignment of semantic values, but it is far from clear which go with which. (See Lewis [9].) There are conflicting tendencies in past usage, and presuppositions we ought to abandon. So I have thought it best to use a newish and neutral term, thereby dodging all issues about which possible sorts of semantic values would deserve which of the familiar names.

Often the truth (-in-English) of a sentence in a context depends on the truth of some related sentence when some feature of the original context is shifted. 'There have been dogs.' is true now iff 'There are dogs.' is true at some time before now. 'Somewhere the sun is shining.' is true here iff 'The sun is shining.' is true somewhere. 'Aunts must be women.' is true at our world iff 'Aunts are women.' is true at all worlds. 'Strictly speaking, France is not hexagonal.' is true even under low standards of precision iff 'France is not hexagonal.' is true under stricter standards.

In such a case, it may be good strategy for a compositional grammar to parse one sentence as the result of applying a modifier to another:

'There have been dogs.' = 'It has been that . . .' + 'There are dogs.'

'Somewhere the sun is shining.' = 'Somewhere . . .' + 'The sun is shining.'

'Aunts must be women.' = 'It must be that . . .' + 'Aunts are women.'

'Strictly speaking, France is not hexagonal.' = 'Strictly speaking . . .' + 'France is not hexagonal.'

Then if the semantic value of the first sentence is to determine its truth in various contexts, and if that value is to be determined by the values of constituents, then the value of the second sentence must provide information about how the second sentence varies in truth value when the relevant feature of context is shifted.

I emphasized that context-dependence was multifarious, but perhaps the shifty kind of context-dependence is less so. The list of shiftable features of context may be quite short. I have suggested that the list should include time, place, world, and (some aspects of) standards of precision. I am not sure what more should be added.

To be sure, we could speak a language in which 'As for you, I am hungry.' is true iff 'I am hungry.' is true when the role of speaker is shifted from me to you – in other words, iff you are hungry. We could – but we don't. For English, the speaker is not a shiftable feature of context. We could speak a language in which 'Backward, that one costs too much.' is true iff 'That one costs too much.' is true under a reversal of the direction the speaker's finger points. But we don't. We could speak a language in which 'Upside down, Fred came floating up through the hatch of the spaceship and turned left.' is true iff 'Fred came floating up through the hatch of the spaceship and turned left.' is true under a reversal of the orientation established in the original context. But we don't. There are ever so many conceivable forms of shiftiness that we don't indulge in.

(To forestall confusion, let me say that in calling a feature of context unshiftable, I do not mean that we cannot change it. I just mean that it does not figure in any rules relating truth of one sentence in context to truth of a second sentence when some feature of the original context is shifted. The established orientation of a context is changeable but probably not shiftable. The world of a context is shiftable but not changeable.)

We seem to have a happy coincidence. To do their first job of determining whether truth-in-English would be achieved if a given sentence were uttered in a given context, it seems that the semantic values of sentences must provide information about the dependence of truth on features of context. That seems to be the very information that is also needed, in view of shiftiness, if semantic values are to do their second job of helping to determine the semantic values of sentences with a given sentence as constituent. How nice.

No; we shall see that matters are more complicated.

6. CONTEXT AND INDEX

Whenever a sentence is said, it is said at some particular time, place, and world. The production of a token is located, both in physical space-time and in logical space. I call such a location a *context*.

That is not to say that the only features of context are time, place, and world. There are countless other features, but they do not vary independently. They are given by the intrinsic and relational character of the time, place, and world in question. The speaker of the context is the one who is speaking at that time, at that place, at that world. (There may be none; not every context is a context of utterance. I here ignore the possibility that more than one speaker might be speaking at the same time, place, and world.) The audience, the standards of precision, the salience relations, the presuppositions . . . of the context are given less directly. They are determined, so far as they are determined at all, by such things as the previous course of the conversation that is still going on at the context, the states of mind of the participants, and the conspicuous aspects of their surroundings.

Suppose a grammar assigns semantic values in such a way as to determine, for each context and each sentence (or for each disambiguation of each sentence), whether that sentence is true in that context. Is that enough? What more could we wish to know about the dependence of truth on features of context?

That is not enough. Unless our grammar explains away all seeming cases of shiftiness, we need to know what happens to the truth values of constituent sentences when one feature of context is shifted and the rest are held fixed. But features of context do not vary independently. No two contexts differ by only one feature. Shift one feature only, and the result of the shift is not a context at all.

Example: under one disambiguation, 'If someone is speaking here then I exist.' is true at any context whatever. No shift from one context to another can make it false. But a time shift, holding other features fixed, can make it false; that is why 'Forevermore, if someone is speaking here then I will exist.' is false in the original context. Likewise a world shift can make it false; that is why 'Necessarily, if someone is speaking here then I must exist.' is false in the original context. The shifts that make the sentence false must not be shifts from one context to another.

The proper treatment of shiftiness requires not contexts but *indices*: packages of features of context so combined that they *can* vary independently. An index is an n-tuple of features of context of various sorts; call these features the *coordinates* of the index. We impose no requirement that the coordinates of an index should all be features of any one context. For instance, an index might have among its coordinates a speaker, a time before his birth, and a world where he never lived at all. Any n-tuple of things of the right kinds is an index. So, although we can never go from one context to another by shifting only one feature, we can always go from one index to another by shifting only one coordinate.

Given a context, there is an index having coordinates that match the appropriate features of that context. Call it the *index of* the context. If we start with the index of a context and shift one coordinate, often the result will be an index that is not the index of any context. That was the case for the time shifts and world shifts that made our example sentence 'If someone is speaking here then I exist.' go false.

Contexts have countless features. Not so for indices: they have the features of context that are packed into them as coordinates, and no others. Given an index, we cannot expect to recover the salience relations (for example) by asking what was salient to the speaker of the index at the time of the index at the world of the index. That method works for a context, or for the index of a context, but not for indices generally. What do we do if the speaker of the index does not exist at that time at that world? Or if the speaker never exists at that world? Or if the time does not exist at the world, since that world is one with circular time? The only way we can recover salience relations from an arbitrary index is if we have put them there as coordinates, varying independently of the other coordinates. Likewise for any other feature of context.

I emphasized that the dependence of truth on context was surprisingly multifarious. It would be no easy matter to devise a list of all the features of context that are sometimes relevant to truth-in-English. In [7] I gave a list that was long for its day, but not nearly long enough. Cresswell rightly complained:

Writers who, like David Lewis, . . . try to give a bit more body to these notions talk about times, places, speakers, hearers, . . . etc. and then go through agonies of conscience in trying to decide whether they have taken account of enough. It seems to me impossible to lay down in advance what sort of thing is going to count [as a relevant feature of context]. . . . The moral here seems to be that there is no way of specifying a finite list of contextual coordinates. ([2], p. 8)

Cresswell goes on to employ objects which, though not the same as the time–place–world locations I have called contexts, are like them and unlike indices in giving information about indefinitely many features of context.

To do their first job of determining whether truth-in-English would be achieved if a given sentence were uttered in a given context, the semantic values of sentences must provide information about the dependence of truth on context. Dependence on indices won't do, unless they are built inclusively enough to include every feature that is ever relevant to truth. We have almost certainly overlooked a great many features. So for the present, while the task of constructing an explicit grammar is still unfinished, the indices we know how to construct won't do. Indices are no substitute for contexts because contexts are rich in features and indices are poor.

To do their second job of helping to determine the semantic values of sentences with a given sentence as a constituent, the semantic values of sentences must provide information about the dependence of truth on indices. Dependence on contexts won't do, since we must look at the variation of truth value under shifts of one feature only. Contexts are no substitute for indices because contexts are not amenable to shifting.

Contexts and indices will not do each other's work. Therefore we need both. An adequate assignment of semantic values must capture two different dependencies of truth on features of context: context-dependence and index-dependence. We need the relation: sentence s is true at context c at index i. We need both the case in which i is the index of the context c and the case in which i has been shifted away, in one or more coordinates, from the index of the context. The former case can be abbreviated. Let us say that sentence s is true at context c iff s is true at c at the index of the context c.

Once we help ourselves to contexts and indices both, we need not go through agonies of conscience to make sure that no relevant features of context have been left out of the coordinates of our indices. Such difficult inclusiveness is needed only if indices are meant to replace contexts. If not, then it is enough to make sure that every shiftable feature of context is included as a coordinate. If most features of context that are relevant to truth are unshiftable, as it seems reasonable to hope, then it might not be too hard to list all the shiftable ones.

8. SCHMENTENCES

Besides the ambitious plan of dispensing with contexts after learning how to construct sufficiently inclusive indices, there is another way to evade my conclusion that we need context-dependence and index-dependence both. The latter was needed only for the treatment of shiftiness, and we might claim that there is no such thing. We can perfectly well build a compositional grammar in which it never happens that sentences are constituents of other sentences, or of anything else. (Make an exception if you like for truth-functional compounding, which isn't shifty; but I shall consider the strategy in its most extreme form.) In this grammar sentences are the output, but never an intermediate step, of the compositional process.

If we take this course, we will need replacements for the sentences hitherto regarded as constituents. The stand-ins will have to be more or less sentence-like. But we will no longer call them sentences, reserving that title for the output sentences. Let us call them *schmentences* instead. We can go on parsing 'There have been dogs.' as the result of applying 'It has been that . . .' to 'There are dogs.'; but we must now distinguish the constituent *schmentence* 'There are dogs.' from the homonymous *sentence,* which is not a constituent of anything. Now the semantic values of genuine sentences have only the first of their former jobs: determining whether truth-in-English would be achieved if a given sentence were uttered in a given context. For that job, dependence of truth on context is all we need. The second job, that of helping to determine the semantic values of sentences with a given constituent, now belongs to the semantic values of schmentences. That job, of course, still requires index-dependence (and context-dependence too, unless the indices are inclusive enough). But nothing requires index-dependent truth of genuine sentences. Instead of giving the semantic values of sentences what it takes to do a double job, we can divide the labour.

32

A variant of the schmentencite strategy is to distinguish schmentences from sentences syntactically. We might write the schmentences without a capital letter and a period. Or we might decorate the schmentences with free variables as appropriate. Then we might parse 'There have been dogs.' as the result of applying 'It has been that . . .' to the schmentence 'there are dogs at t' where 't' is regarded as a variable over times. The confusing homonymy between schmentences and sentences is thereby removed. Index-dependence of the schmentence thus derives from index-dependence of the values of its variables. Schmentences would be akin to the open formulas that figure in the standard treatment of quantification. Truth of a schmentence at an index would be like satisfaction of a formula by an assignment of values to variables. But while the schmentencite might proceed in this way, I insist that he need not. Not all is a variable that varies. If the coordinates of indices were homogeneous in kind and unlimited in number – which they are not – then it might be handy to use variables as a device for keeping track of exactly how the truth value of a schmentence depends on the various coordinates. But variables can be explained away even then (see Quine [14]); or rather, they can be replaced by other devices to serve the same purpose. If the coordinates of indices are few and different in kind, it is not clear that variables would even be a convenience.

(Just as we can liken index-dependent schmentences to formulas that depend for truth on the assignment of values to their free variables, so also we can go in the reverse direction. We can include the value assignments as coordinates of indices, as I did in [7], and thereby subsume assignment-dependence of formulas under index-dependence of sentences. However, this treatment is possible only if we limit the values of variables. For instance we cannot let a variable take as its value a function from indices, since that would mean that some index was a member of a member of . . . a member of itself – which is impossible.)

I concede this victory to the schmentencite: strictly speaking, we *do not need* to provide both context-dependence and index-dependence in the assignment of semantic values to genuine sentences. His victory is both cheap and pointless. I propose to ignore it.

Therefore, let us agree that sentences depend for their truth on both context and index. What, then, should we take as their semantic values? We have two options.

First option: the semantic values of sentences are variable but simple. A value for a sentence is a function, perhaps partial, from indices to truth values. (Alternatively, it is a set of indices.) However, a sentence may have different semantic values in different contexts, and the grammar must tell us how value depends on context. The grammar assigns a semantic value (or more than one, in case of ambiguity) to each sentence-context pair. The value in turn is something which, together with an index, yields a truth value. Diagrammatically:

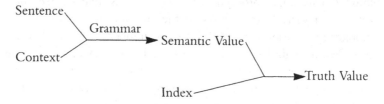

Sentence s is true at context c at index i iff $V_c^s(i)$ is truth, where V_c^s is the value of s at c. Sentence s is true at context c iff $V_c^s(i_c)$ is truth, where i_c is the index of the context c.

Second option: the semantic values of sentences are *constant but complicated*. A value for a sentence is a function, perhaps partial, from combinations of a context and an index to truth values. (Alternatively, it is a set of context-index combinations.) The semantic value of a sentence (or its set of values, in case of ambiguity) does not vary from context to context. The grammar assigns it once and for all. Diagrammatically:

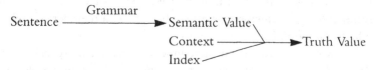

Sentence s is true at context c at index i iff $V^s(c + i)$ is truth, where V^s is the constant semantic value of s. Sentence s is true at context c

iff $V^s(c + i_c)$ is truth, where i_c is the index of the context c. Context-index combinations could be taken in either of two ways: as pairs $\langle c, i \rangle$ of a context c and an index i, or alternatively as $(n + 1)$-tuples $\langle c, i_1, \ldots, i_n \rangle$ that start with c and continue with the n coordinates of i.

(It is worth mentioning and rejecting a zeroth option: the semantic values of sentences are *very variable but very simple*. They are simply truth values; however, a sentence has different semantic values at different context-index combinations. This option flouts the compositional principle, which requires that the semantic values of sentences be determined by the semantic values of their constituent sentences. The truth value of a sentence at a given context and index may depend not on the truth value of a constituent sentence at that context and index, but rather on its truth value at that context and other, shifted indices. The less I have said about what so-called semantic values must be, the more I am entitled to insist on what I *did* say. If they don't obey the compositional principle, they are not what I call semantic values.)

Asked to choose between our two options, you may well suspect that we have a distinction without a difference. Given a grammar that assigns semantic values according to one option, it is perfectly automatic to convert it into one of the other sort. Suppose given a grammar that assigns variable but simple semantic values: for any sentence s and context c, the value of s at c is V_c^s. Suppose you would prefer a grammar that assigns constant but complicated values. Very well: to each sentence s, assign once and for all the function V^s such that, for every context c and index i, $V^s(c + i)$ is $V_c^s(i)$. Or suppose given a grammar that assigns constant but complicated semantic values: to sentence s it assigns, once and for all, the value V^s. Suppose you would prefer a grammar that assigns variable but simple values. Very well: to the sentence s and context c, assign the function V_c^s such that, for every index i, $V_c^s(i)$ is $V^s(c + i)$.

Given the ease of conversion, how could anything of importance possibly turn on the choice between our two options? Whichever sort of assignment we are given, we have the other as well; and the assigned entities equally well deserve the name of semantic values because they equally well do the jobs of semantic values. (If we asked whether they equally well deserved some other name in our traditional semantic vocabulary, that would be a harder question but an idle one. If we asked whether they would fit equally well into future psycholinguistics, that would – in my opinion – be a question so hard and speculative as to be altogether futile.) How could the choice between the options possibly be a serious issue?

I have certainly not taken the issue very seriously. In [7] I opted for constant but complicated semantic values (though not quite as I described them here, since I underestimated the agonies of constructing sufficiently rich indices). But in [6] and [8], written at about the same time, I thought it made for smoother exposition to use variable but simple values (again, not quite as described here). I thought the choice a matter of indifference, and took for granted that my readers would think so to.

But I was wrong. Robert Stalnaker [11] and David Kaplan [5] have taken the issue very seriously indeed. They have argued that we ought to prefer the first option: variable but simple semantic values. Each thinks that simple, context-dependent semantic values of the proper sort (but not complicated constant ones) are good because they can do an extra job, besides the two jobs for semantic values that we have been considering so far. They differ about what this extra job is, however, and accordingly they advocate somewhat different choices of variable but simple values.

10. CONTENT AS OBJECT OF ATTITUDES: STALNAKER

In Stalnaker's theory, the semantic value of a sentence in context (after disambiguation) is a *proposition:* a function from possible worlds to truth values. Diagrammatically:

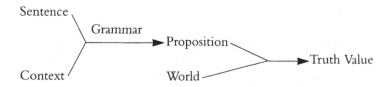

He mentions the alternative analysis on which a sentence is assigned, once and for all, a function from context-world combinations to truth values.

It is a simpler analysis than the one I am sketching; I need some argument for the necessity or desirability of the extra step on the road from sentences to truth values. This step is justified only if the middlemen – the propositions – are of some independent interest,. . . . The independent interest in propositions comes from the fact that they are the objects of illocutionary acts and propositional attitudes. A proposition is supposed to be the common content of statements, judgements, promises, wishes and wants, questions and answers, things that are possible or probable. ([11], pp. 277–278)

I agree with much of this. Stalnaker is right that we can assign propositional content to sentences in context, taking propositions as functions from worlds to truth values. He is also right that propositions have an independent interest as suitable objects for attitudes such as belief, and in the other ways he mentions. (Here I pass over a big idealization; it could be defended in several ways and I am not sure which I prefer.) Furthermore, an account of truthful communication – not part of the grammar itself, but another chapter in the systematic restatement of our common knowledge about language – must concern itself at least implicitly with the relations between the propositional objects of the speaker's attitudes and the propositional content of his sentences.

To revert to our initial example: I know, and you need to know, whether A or B or . . . ; so I say a sentence that I take to be true-in-English, in its context, and that depends for its truth on whether A or B or . . . ; and thereby, if all goes well, you find out what you needed to know. My choice of what to say is guided by my beliefs. It depends on whether I believe the proposition true at exactly the A-worlds, or the one true at exactly the B-worlds, or. . . . In the simplest case, the sentence I choose to say is one whose propositional content (in English, in context) is whichever one of these propositions I believe.

That is all very well, but it does not mean that we need to equate the propositional content and the semantic value of a sentence in context. It is enough that the assignment of semantic values should somehow determine the assignment of propositional content. And it does, whether we opt for variable but simple values or for constant but complicated ones. Either way, we have the relation: sentence s is true at context c at index i. From that we can define the propositional content of sentence s in context c as that proposition that is true at world w iff s is true at c at the index i_c^w that results if we take the index i_c of the context c and shift its world coordinate to w.

(We can call this the *horizontal propositional content* of s in c; borrowing and modifying a suggestion of Stalnaker in [12] we could also define the *diagonal propositional content* of s in c. Suppose someone utters s in c but without knowing whether the context of his utterance is c or whether it is some other possible context in some other world which is indistinguishable from c. Since all ignorance about contingent matters of fact is ignorance about features of context, the sort of ignorance under consideration is a universal predicament. Let c^w be that context, if there is one, that is located at world w and indistinguishable from c; then for all the speaker knows he might inhabit w and c^w might be the context of his utterance. (I ignore the case of two indistinguishable contexts at the same world.) Let $i_c w$ be the index of the context c^w; note that this may differ from the index i_c^w mentioned above, since the contexts c and c_w will differ not only in world but in other features as well and the indices of the differing contexts may inherit some of their differences. We define the diagonal content of s in c as that proposition that is true at a world w iff (1) there is a context c^w of the sort just considered, and (2) s is true at c^w at $i_c w$. Stalnaker shows in [12] that horizontal and diagonal content both enter into an account of linguistic communication. The former plays the central role if there is no significant ignorance of features of context relevant to truth; otherwise we do well to consider the latter instead. Stalnaker speaks of reinterpreting sentences in certain contexts so that they express their diagonal rather than their horizontal content. I find this an inadvisable way of putting the point, since if there is a horizontal-diagonal ambiguity it is very unlike ordinary sorts of ambiguity. I doubt that we can perceive it as an ambiguity; it is neither syntactic nor lexical; and it is remarkably widespread. I think it might be better to say that a sentence in context has both a horizontal and a diagonal content; that these may or may not be the same; and that they enter in different ways into an account of communication. Be that as it may, I shall from now on confine my attention to propositional content of the horizontal sort; but what I say would go for diagonal content also.)

It would be a convenience, nothing more, if we could take the propositional content of a sentence in context as its semantic value. But we cannot. The propositional contents of sentences do not obey the compositional principle, therefore they are not semantic values. Such are the ways of shiftiness that the propositional content of 'Somewhere the sun is shining.' in context c is not determined by the content in c of the constituent sentence 'The sun is shining.'. For an adequate treatment of shiftiness we need not just world-dependence but index-dependence – dependence of truth on all the shiftable features of context. World is not the only shiftable feature.

(Stalnaker does suggest, at one point, that he might put world-time pairs in place of worlds. "Does a tensed sentence determine a proposition which is sometimes true, sometimes false, or does it express different timeless propositions at different times? I doubt that a single general answer can be given." ([11], p. 289) But this does not go far enough. World and time are not the only shiftable features of context. And also perhaps it goes too far. If propositions are reconstrued so that they may vary in truth from one time to another, are they still suitable objects for propositional attitudes?*)

There is always the schementencite way out: to rescue a generalization, reclassify the exceptions. If we said that the seeming sentences involved in shiftiness of features other than world (and perhaps time) were not genuine sentences, then we would be free to say that the semantic value of a genuine sentence, in context, was its propositional content. But what's the point?

I have been a bit unfair to complain that the propositional content of a sentence in context is not its semantic value. Stalnaker never said it was. 'Semantic value' is my term, not his. Nor did he falsely claim that contents obeys the compositional principle.

But my point can be stated fairly. Nothing is wrong with what Stalnaker says, but by omission he gives a misleading impression of simplicity. Besides the propositional content of a given sentence in a given context, and besides the function that yields the content of a given sentence in any context, we need something more – some-

* [Added 1996] Yes indeed. For discussion, see my 'Attitudes *De Dicto* and *De Se*', *The Philosophical Review* 88 (1979), pp. 513–543.

thing that goes unmentioned in Stalnaker's theory. We need an assignment of semantic values to sentences (or to schmentences) that captures the dependence of truth both on context and on index, and that obeys the compositional principle. An assignment of variable but simple semantic values would meet the need, and so would an assignment of constant but complicated ones. Neither of these could *be* the assignment of propositional content. Either would suffice to determine it. So Stalnaker's discussion of propositional content affords no reason for us to prefer variable but simple semantic values rather than constant but complicated ones.

11. CONTENT AS WHAT IS SAID: KAPLAN

Kaplan [5], unlike Stalnaker, clearly advocates the assignment of variable but simple semantic values as I have described it here. His terminology is like Stalnaker's, but what he calls the content of a sentence in context is a function from moderately rich indices to truth values. Diagrammatically:

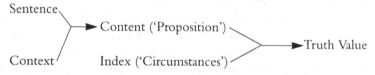

I cannot complain against Kaplan, as I did against Stalnaker, that his so-called contents are not semantic values because they violate compositionality. But Kaplan cannot plausibly claim, as Stalnaker did, that his contents have an independent interest as suitable objects for propositional attitudes.

Kaplan claims a different sort of independent interest for his contents – that is, for variable but simple semantic values. We have the intuitive, pre-theoretical notion of 'what is said' by a sentence in context. We have two sentences in two contexts, or one sentence in two contexts, or two sentences in one context; and we judge that what has been said is or is not the same for both sentence-context pairs. Kaplan thinks that if we assign simple, context-dependent semantic values of the right sort, then we can use them to explicate our judgements of sameness of what is said: what is said by sentence

s_1 in context c_1 is the same as what is said by sentence s_2 in context c_2 iff the semantic value of s_1 in c_1 and the semantic value of s_2 in c_2 are identical. Indeed, Kaplan suggests that our informal locution 'what is said' is just a handy synonym for his technical term 'content'.

Thus if I say, today, 'I was insulted yesterday.' and you utter the same words tomorrow what is said is different. If what we say differs in truth value, that is enough to show that we say different things. But even if the truth values were the same, it is clear that there are possible circumstances in which what I said would be true but what you said would be false. Thus we say different things. Let us call this first kind of meaning – what is said – *content*. ([5], p. 19)

Consider some further examples. (1) I say 'I am hungry.'. You simultaneously say to me 'You are hungry.'. What is said is the same. (2) I say 'I am hungry.'. You simultaneously say 'I am hungry.'. What is said is not the same. Perhaps what I said is true but what you said isn't. (3) I say on 6 June 1977 'Today is Monday.'. You say on 7 June 1977 'Yesterday was Monday.'. What is said is the same. (4) Same for me, but you say on 7 June 1977 'Today is Tuesday.'. What is said is the same. (5) I say on 6 June 1977 'It is Monday.'. I might have said, in the very same context, '6 June 1977 is Monday.'. or perhaps 'Today is Monday.'. What is said is not the same. What I did say is false on six days out of every seven, whereas the two things I might have said are never false.

I put it to you that not one of these examples carries conviction. In every case, the proper naive response is that in some sense what is said is the same for both sentence-context pairs, whereas in another – equally legitimate – sense, what is said is not the same. Unless we give it some special technical meaning, the locution 'what is said' is very far from univocal. It can mean the propositional content, in Stalnaker's sense (horizontal or diagonal). It can mean the exact words. I suspect that it can mean almost anything in between. True, what is said is the same, in some sense, iff the semantic value is the same according to a grammar that assigns variable but simple values. So what, unless the sense in question is more than one among many? I think it is also so that what is said is the same, in some sense, iff the semantic value is the same according to a grammar that assigns constant but complicated values.

Kaplan's readers learn to focus on the sense of 'what is said' that he has in mind, ignoring the fact that the same words can be used to make different distinctions. For the time being, the words mark a definite distinction. But why mark that distinction rather than others that we could equally well attend to? It is not a special advantage of variable but simple semantic values that they can easily be used to explicate those distinctions that they can easily be used to explicate.

12. SOLIDARITY FOREVER

I see Stalnaker and Kaplan as putting forth package deals. Offered the whole of either package – take it or leave it – I take it. But I would rather divide the issues. Part of each package is a preference, which I oppose as unwarranted and arbitrary, for variable but simple semantic values. But there is much in each package that I applaud; and that I have incorporated into the proposals of the present paper, whichever option is chosen. In particular there are three points on which Stalnaker and Kaplan and I join in disagreeing with my earlier self, the author of [7].

First, the author of [7] thought it an easy thing to construct indices richly enough to include all features of context that are ever relevant to truth. Stalnaker and Kaplan and I all have recourse to genuine context-dependence and thereby shirk the quest for rich indices. Stalnaker and Kaplan do not dwell on this as a virtue of their theories, but it is one all the same.

Second, I take it that Stalnaker and Kaplan and I join in opposing any proposal for constant but complicated but not complicated enough semantic values that would ignore the following distinction. There are sentences that are true in any context, but perhaps not necessarily true; and there are sentences in context that are necessarily true, though perhaps the same sentence is not necessarily true, or not true at all, in another context. (This is at least an aspect of Kripke's well-known distinction between the *a priori* and the necessary.) The distinction might be missed by a treatment that simply assigns functions from indices to truth values (as in [7]), or functions from contexts to truth values, as the constant semantic values of sentences. It is captured by any treatment that combines context-dependence and index-dependence, as in Kaplan's theory or the treatment proposed here; it is likewise captured by any treatment that combines context-dependence and world-dependence, as in Stalnaker's theory or my [6] and [8]. In the first case it is the distinction between (1) a sentence that is true at every context c at the index i of c, and (2) a sentence that is true at a particular context c at every index i^w that comes from the index i of the context c by shifting the world coordinate. In the second case it is the distinction between (1) a sentence that is true at every context c at the world of c, and (2) a sentence that is true at some particular context c at every world.

Third, all three of us, unlike the author of [7], have availed ourselves of the device of *double indexing*. Context-dependence and index-dependence (or world-dependence) together give a double world-dependence: truth may depend both on the world of the context and on the world-coordinate of the index, and these may differ since the latter may have been shifted. That facilitates the semantic analysis of such modifiers as 'actually': 'Actually ϕ.' is true at context c at index i iff ϕ is true at c at i^w, the index that comes from i by shifting the world coordinate to the world w of the context c. Similarly, context-dependence and index-dependence together give a double time-dependence (if indices have time coordinates) so that we can give a version of Kamp's analysis of 'now': 'Now ϕ.' is true at context c at index i iff ϕ is true at c at i^t, the index that comes from i by shifting the time coordinate to the time t of the context c.

For extensive discussions of the uses and origins of double indexing, see Kaplan [5] and van Fraassen [13]. However, there is a measure of disappointment in store. For some uses of double indexing, it is enough to have double world-dependence (or time-dependence) in which the world (or time) appears once shiftably and once unshiftably. 'Actually' (or 'now'), for instance, will always bring us back to the world (or time) of the context. For these uses, the extra world-dependence and time-dependence that come as part of context-dependence will meet our needs. But there are other applications of double indexing, no less useful in the semanticist's toolbox, that require double shiftability. The principal application in [13] is of this sort. Moreover, if we combine several applications that each require double shiftability, we may then need more than double indexing. Coordinates that have been shifted for one purpose are not available unshifted for another purpose. If we want multiply shiftable multiple indexing, then we will have to repeat the world or time co-ordinates of our indices as many times over as needed. The unshiftable world and time of the context will take us only part of the way.

BIBLIOGRAPHY

[1] Yehoshua Bar-Hillel, 'Indexical Expressions', *Mind* **63** (1954), 359–379.

[2] M. J. Cresswell, 'The World is Everything That is the Case', *Australasian Journal of Philosophy* **50** (1972), 1–13.

[3] Charles Fillmore, 'How to Know Whether You're Coming or Going', in *Studies in Descriptive and Applied Linguistics: Bulletin of the Summer Institute in Linguistics V* (International Christian University, Tokyo).

[4] Robin Cooper and Terence Parsons, 'Montague Grammar, Generative Semantics and Interpretive Semantics', in Barbara Partee (ed.), *Montague Grammar* (Academic Press, 1976).

[5] David Kaplan, 'Demonstratives', presented at the 1977 meeting of the Pacific Division of the American Philosophical Association. (Precursors of this paper, with various titles, have circulated widely since about 1971.)

[6] David Lewis, *Convention: A Philosophical Study* (Harvard University Press, 1969).

[7] David Lewis, 'General Semantics', *Synthese* **22** (1970), 18–67; and in Barbara Partee (ed.), *Montague Grammar* (Academic Press, 1976).

[8] David Lewis, 'Languages and Language', in *Minnesota Studies in the Philosophy of Science,* Vol. VII (University of Minnesota Press, 1975).

[9] David Lewis, ''Tensions', in Milton Munitz and Peter Unger (eds.), *Semantics and Philosophy* (New York University Press, 1974).

[10] Richard Montague, 'Pragmatics', in Montague, *Formal Philosophy* (Yale University Press, 1974).

[11] Robert Stalnaker, 'Pragmatics', *Synthese* **22** (1970), 272–289.

[12] Robert Stalnaker, 'Assertion', in Peter Cole (ed.), *Syntax and Semantics 9* (New York, 1978).

[13] Bas van Fraassen, 'The Only Necessity is Verbal Necessity', *Journal of Philosophy* **74** (1977), 71–85.

[14] W. V. Quine, 'Variables Explained Away', in Quine, *Selected Logic Papers* (Random House, 1966).

3

'Whether' report

KNOWING WHETHER AND TELLING WHETHER

Mr. Body lies foully murdered, and the suspects are Green, Mustard, Peacock, Plum, Scarlet, and White. We may take it as settled that one of them did it, and only one. The question is whether Green did it, or Mustard did it, or Peacock, or Plum, or Scarlet, or White. Holmes is on the scene.

If Green did it, then Holmes knows whether Green did it or . . . or White did it if and only if he knows that Green did it. Likewise if Mustard did it, then Holmes knows whether . . . if and only if he knows that Mustard did it. Likewise for the other cases. In short, Holmes knows whether . . . if and only if he knows the true one of the alternatives presented by the 'whether'-clause, whichever one that is.

Similarly for telling. In at least one principal sense, Holmes tells Watson whether Green did it, or Mustard did it, or Peacock, or Plum, or Scarlet, or White, if and only if Holmes tells Watson the true one of the alternatives presented by the 'whether'-clause. That is: if and only if either Green did it and Holmes tells Watson that

First published in Tom Pauli *et al.* (eds.) *320311: Philosophical Essays Dedicated to Lennart Åqvist on his Fiftieth Birthday* (Uppsala, Filosofiska Studier, 1982). Reprinted with kind permission from Filosofiska Studier.

An ancestor of this paper was written in 1974 as a contribution to a Workshop on Semantics and Syntax of Non-Extensional Constructions, held at the University of Massachusetts, Amherst, and sponsored by the Mathematical Social Sciences Board. I am grateful to the participants for their comments, to the University for hospitality, and to the Board for support.

Green did it, or . . . or White did it and Holmes tells Watson that White did it.

This is a *veridical* sense of telling whether, in which telling falsely whether does not count as telling whether at all, but only as purporting to tell whether. This veridical sense may or may not be the only sense of *'tell whether'*; it seems at least the most natural sense.

'Whether'-clauses may be abbreviated. Holmes knows or tells whether Green or . . . or White did it if and only if he knows or tells whether Green did it or . . . or White did it. He knows or tells whether Green did it or not if and only if he knows or tells whether Green did it or Green did not do it. He knows or tells whether Green did it if and only if he knows or tells whether Green did it or not. And of course we may also abbreviate by putting plain commas in place of all but the last of the *'or'*s.

IGNORANCE, SECRECY, WONDERING, AND ASKING WHETHER

Some other constructions with 'whether' may be analyzed in terms of knowing or telling whether. Lestrade's ignorance as to whether Green did it or . . . or White did it consists in his not knowing whether Green did it or . . . or White did it. He remains ignorant because Holmes keeps it secret from him whether Green did it or . . . or White did it, and Holmes does so exactly by knowing whether while declining to tell him whether. Consequently Lestrade wonders whether Green did it or . . . or White did it; that is (1) he does not know whether Green did it or . . . or White did it; (2) he desires an end to this ignorance; and (3) this desire is part of his conscious thought.

When Holmes told Watson whether Green did it or . . . or White did it, that was because Watson had asked him. Watson had requested that Holmes tell him whether. Or perhaps Watson, knowing Holmes' distaste for the straight answer, rather had requested that Holmes see to it somehow that Watson come to know whether.

In terms of Åqvist's imperative-epistemic analysis of questions, given in [6], [7], [8], and [9], Watson's question to Holmes could be formalized as

Let it be that: you see to it that: I soon come to know: whether (Green did it, . . . , White did it)

where this ought to come out equivalent to

Let it be that: you see to it that: (I soon come to know that Green did it ∨ . . . ∨ I soon come to know that White did it).

Three comments. First, we need not make it explicit that the one to become known is to be the true one, for only truth can possibly become known. Second, the operator of agency *'you see to it that'* and the adverb *'soon'* are not present, or at any rate not explicit, in Åqvist's own formulation; but, as noted in [2], they seem to be desirable – though complicating – additions. Third, we need to do something to stop the implication to

Let it be that: you see to it that: (Green did it ∨ . . . ∨ White did it),

or else to reinterpret this consequence so that it does not amount to a request on Watson's part that Holmes bring about a past murder. The problem may have various solutions; Åqvist presents one solution (credited to Hintikka) in [9].

A close relative of Åqvist's analysis is the imperative-assertoric analysis discussed *inter alia* in [2]. It seems preferable in many cases, though perhaps (given that Watson takes account of Holmes' penchant for making things known otherwise than by telling them) the original imperative-epistemic analysis better suits our present case. Be that as it may, an imperative-assertoric analysis of Watson's question could run as follows.

Let it be that: you tell me: whether (Green did it, . . . , White did it)

where this ought to come out equivalent to

Let it be that: ((Green did it & you tell me that Green did it) ∨ . . . ∨ (White did it & you tell me that White did it)).

Two of our previous comments apply, *mutatis mutandis*. First, the request imputed to Watson on this analysis is again somewhat redundant. He could simply have requested that Holmes tell him some one of the alternatives, without requesting that Holmes pick the true one. Indeed Holmes *could* have told him one of the others; but Holmes *would* not have done so, being an honourable man dealing with a close friend. Watson was in a position to rely on Holmes' truthfulness without especially requesting it – indeed, had he not been, neither could he have relied on Holmes to be truthful on request. The same is true generally: those questioned are supposed to tell the truth without any special request. (On this point I am indebted to conversation with John Searle.) But it is harmless to impute to Watson a needlessly strong request; and I do so in order to connect the question '*Tell me whether . . . ?*' with the veridical sense of '*tell whether*'. Second, we need to do something to stop the implication to

Let it be that: (Green did it ∨ *. . .* ∨ *White did it),*

or else we need to reinterpret this conclusion so that it is not a malevolent optative.

In short: Watson's question was a request that Holmes tell him, or at any rate that Holmes somehow make known to him, the true one of the alternatives presented by the 'whether'-clause.

STRATEGIES FOR SEMANTIC ANALYSIS

Suppose that we wished to provide a formal semantic analysis for a fragment of English that includes the constructions just considered: constructions of knowing and telling whether, and other constructions analyzable in terms of these. How might we treat 'whether'-clauses? At least five alternative strategies come to mind.

A. *We might eliminate 'whether'-clauses altogether.* Rather than taking '*whether*' plus its appended list of alternatives as a constituent, we might take '*whether*' as part of the verb. Thus '*know whether*', '*tell whether*', and '*ask whether*', for instance, would be verbs that attach to lists of sentences and in some cases also to indirect objects to make intransitive verb phrases. These verbs would be semantically indivisible primitives: their semantic values would not be derived from the semantic values of '*know*', '*tell*', and '*ask*', nor would the word '*whether*' have any semantic value of its own. Hence this strategy passes up opportunities for unification both in its treatment of different constructions with '*know*' et al. and in its treatment of different constructions with '*whether*'.

B. *We might assign 'whether'-clauses to a special category all their own,* or to a special category reserved for wh-clauses generally. This would at least permit a unified treatment of 'whether'; but the treatment of '*know*' et al. remains disunified, since we must distinguish the 'know' that attaches to a 'whether'-clause from the categorically different '*know*' of '*Holmes knows that Peacock is innocent*'. And, other things being equal, we should not multiply categories.

C. *We might treat 'whether'-clauses as terms denoting their sets of alternatives.* Now we have a unified treatment of '*whether*' and we need no special category. But we still have a disunified treatment of '*know*': sometimes the knower is related to a single propositional content (or content of some other appropriate sort), sometimes rather to a set of alternative contents. Likewise for '*tell*'. Also, we have a new problem, avoided by strategies A and B: a term denoting a set of alternative contents ought to make sense in various positions where a 'whether'-clause cannot in fact occur:

* *Whether Green did it or Plum did it has two members.*

D. *We might treat 'whether'-clauses as terms denoting the true one of the alternatives they present.* Now we get full unification, at least for the constructions so far considered. If Green did it, then we may take '*whether Green did it or White did it*' and '*that Green did it*' as two terms denoting the same thing, and capable of being attached to the same verb '*know*' or '*tell*'. But again we need arbitrary-seeming restrictions to stop the occurrence of 'whether'-clauses in positions where terms denoting the true one of the alternatives ought to make sense:

* *Whether Mustard or Scarlet did it is true,*
* *Whether Mustard or Scarlet did it is some sort of abstract entity.*

Of course, the same difficulty arises to some extent with 'that'-clauses regarded as denoting terms:

* *That Plum did it is some sort of abstract entity.*

E. *We might treat 'whether'-clauses as sentences expressing the same content as the true one of the presented alternatives.* If we do not wish to treat 'that'-clauses as denoting terms, perhaps because of such difficulties as the one just noted, we could treat *'know'* and *'tell'* as verbs that attach to sentences. Then the word *'that'* would not be assigned any semantic value, though syntax would still have to take account of it. This strategy imitates those formal languages that contain epistemic or assertoric modal operators. If we do this, and we still seek unification, 'whether'-clauses also must be treated as sentences; and the content of the 'whether'-clauses must be the same as that of the true alternative. We shall see how this is possible.

The problem of arbitrarily prohibited occurrences is still with us, though our previous examples of it are disposed of when we no longer treat 'whether'-clauses as terms. For instance, why can't we say

* *Lestrade believes whether Mustard or Scarlet did it*

to mean that Lestrade believes the true one of the alternatives thereby presented?

Strategies D and E have the advantage of unifying the treatment of *'know'* and *'tell'*; but C, and perhaps also B, may do better at unifying the treatment of *'whether'* itself. For there is a second family of constructions with *'whether'* besides the ones we have considered so far; and in these, what matters is not especially the true one of the alternatives but the whole range of them. For these constructions, the step to denoting (or expressing) the true one of the alternatives is a step in an unhelpful direction. This second family consists of constructions expressing dependence or independence: *'whether . . . or . . . depends on . . . ', '. . . depends on whether . . . or . . . ', 'no matter whether . . . or . . . , still . . . ', 'it doesn't matter whether . . . or . . . '*, and so on.

So the advantages are not all on one side; and I would not wish to resurrect the 'whether' report that I once wrote (and then suppressed) which definitely advocated strategy E. Nevertheless, it remains of interest that strategy E is even possible; and it does have its attractions. Let us examine it further.

For the remainder of this paper, I shall suppose that *'whether'* is a sentential operator of variably many places: it attaches to two or more sentences to make a sentence. The sentences it attaches to are punctuated by *'or's*, perhaps supplemented or replaced at all but the final occurrence by commas.

I shall also suppose that the content expressed by a sentence (in context) is the same sort of thing that can be known or that can be told; and that it is the same thing as the *truth set* of the sentence: the set of maximally specific possibilities that make the sentence true (under the interpretation bestowed on it by its actual context). It is common to take these possibilities as abstract representations of ways the whole world might be; I am inclined to disagree on two counts, taking them rather as concrete possible individuals. But for the sake of neutrality, let me call them simply *points*. A sentence, then, is true at some points, false at others; and the content it expresses is the set of points at which it is true.

We want a 'whether'-clause to express the same content as does the true one of the sentences embedded in it, provided that exactly one of those sentences is true. But how can we provide for that? If A is the true one of A and B, and we assign the same content to *'whether A or B'* as we do to A itself, then that assignment would have been wrong if B had been the true one! The answer is that we must not assign content to 'whether'-clauses once and for all; rather, we must make the content vary, depending on contingent matters of fact.

That means that we must resort to *double indexing,* a technical device invented by Hans Kamp and Frank Vlach in the late 1960's and since used in many ways by many authors. (See, for instance, [1], [3], [4], [5], and [10].) We shall not say that a sentence is true at a point *simpliciter;* rather, a sentence is true at a point relative to a second point, which may or may not be the same as the first. For most sentences, the second point is idle: *normal* sentences satisfy the condition:

$\vdash_{i,j} A$ if and only if $\vdash_{i,k} A$, for any j and k.

It will therefore be useful to reintroduce truth at a single point by way of abbreviation, noting that this will be the only truth relation that matters for normal sentences:

$\mathsf{F}_i A$ is defined as $\mathsf{F}_{i,i} A$.

Content of a sentence relative to a point is then defined as the set of first points that make the sentence true relative to the fixed second point:

$$[\![A]\!]_j = {}^{df}\{i: \mathsf{F}_{i,j} A\}.$$

A normal sentence has the same content relative to all points; an abnormal sentence does not. Sentences made with *'whether'* are, in general, abnormal. They have variable content, as is required by strategy E. Our rule for *'whether'* is as follows.

$\mathsf{F}_{i,j}$ *whether* A_1 *or* . . . *or* A_n
if and only if either $\mathsf{F}_i A_1$ and $\mathsf{F}_j A_1$
or . . . or $\mathsf{F}_i A_n$ and $\mathsf{F}_j A_n$.

If the A's themselves are normal, and if exactly one of them is true at point j, then we have the desired result: the content relative to j of the 'whether'-clause is the same as the content relative to j (or absolutely) of the true one of the A's:

If $\mathsf{F}_j A_1$, $\not\!\mathsf{F}_j A_2$, . . . , $\not\!\mathsf{F}_j A_n$, then $[\![$ *whether* A_1
or . . . *or* $A_n]\!]_j = [\![A_1]\!]_j$;

and so on for the other cases.

Now if we have the natural rules for *'know'* and *'tell'*, so that for instance

F_i *Holmes knows (that)* A if and only if Holmes at point i stands in the knowing-relation to the content $[\![A]\!]_i$

then we will get the right results for knowing whether, telling whether, and whatever is definable in terms of them. Putting our 'whether'-clause in as the sentence A (and relying on our syntax to delete, or not to insert, the word *'that'*) we find as expected that, for instance

If $\vdash_i A_1$, $\nvdash_i A_2, \ldots, \nvdash_i A_n$, then $\vdash_i$ *Holmes knows whether* A_1
or . . . or A_n if and only if Holmes at point i stands in the knowing-relation to the content $[\![A_1]\!]_i$;

and so on for the other cases.

Given our double indexing, it is really not quite enough to give conditions of truth *simpliciter* for sentences such as '*Holmes knows whether* A_1 *or . . . or* A_n,', '*Holmes tells Watson whether* A_1 *or . . . or* A_n,', and their relatives. We must also say when such a sentence is true at a point i relative to a second point j. The simplest stipulation, I think, is that our sentences of knowing and telling are to be normal sentences:

$\vdash_{i,j}$ *Holmes knows (that)* A if and only if
$\vdash_i$ *Holmes knows (that)* A

with the right-hand side evaluated according to the condition of truth *simpliciter* already given. This applies, in particular, to '*Holmes knows whether* A_1 *or . . . or* A_n'. Accordingly, our sentences of knowing whether and telling whether can without trouble be embedded into further 'whether'-clauses, as in

Holmes tells Watson whether or not Lestrade knows whether Plum did it or Green did it.

What happens if, at a certain point j, *none* of the alternatives presented in the 'whether'-clause is true?

If $\nvdash_j A_1, \ldots, \nvdash_j A_n$, then $\nvdash_{i,j}$ *whether* A_1 *or . . . or* A_n

53

so in that case the content relative to j of the 'whether'-clause is the empty set − the content that cannot possibly be true. This impossible content cannot be known; so if none of $A_1, \ldots, A_n$ is true at j, then it cannot be true at j that Holmes knows whether A_1 or . . . or A_n. And that seems exactly right. Also it seems that in this case it cannot be true at j that Holmes tells Watson whether A_1 or . . . or A_n (at any rate, not in our veridical sense of telling whether). This too is a consequence of our rules, provided that the empty set − the impossible content − is something that cannot be told. I think that indeed the empty set cannot be told. (Contradicting oneself should not count as telling one thing with impossible content, but rather as telling two or more things with conflicting contents. Or so I think − but this is highly controversial.) If so, our rule performs as it should in this case too.

What happens if, at a certain point j, *more* than one of the presented alternatives are true? Then I think a presupposition required for the proper use of the 'whether'-clause has failed, with the result that no clear-cut data on truth conditions are available and so we may as well handle the case in the most technically convenient way. The interested reader is invited to discover how I have implicitly handled the case.

What if some of the sentences embedded in the 'whether'-clause are already abnormal, expressing different content relative to different points? Again, the interested reader may discover my implicit answer. But I think the question idle, for there is no reason to believe that the difficulty need ever arise. It would arise if 'whether'-clauses were immediately embedded in other 'whether'-clauses − but that should not be allowed to happen in any correct fragment of English. Perhaps it might arise if our double indexing for 'whether'-clauses interacted with double indexing introduced for some other purpose; but we could always stop such interaction by moving to triple, quadruple, . . . indexing.

'WHETHER'-CLAUSES AS DISJUNCTIONS

Strategy E offers an answer to one question that its rivals do not address. Why the *'or'* in *'whether Green did it or Scarlet did it'*? We need a punctuation mark to separate the listed sentences that give the alternatives, but why do we use 'or' for the job? Why not always commas? Or some special mark just for this purpose? Why does a 'whether'-clause look like a disjunction?

(Of course, we must be careful. We must distinguish 'Holmes knows whether Green did it or Scarlet did it' from 'Holmes knows whether (Green did it or Scarlet did it)'. The latter, with its straightforward disjunction, does not present 'Green did it' and 'Scarlet did it' as separate alternatives. Rather, it abbreviates 'Holmes knows whether (Green did it or Scarlet did it) or not-(Green did it or Scarlet did it)', and the alternatives presented are the disjunction and its negation. If Holmes knew that it was Green or Scarlet but didn't know which, then the latter sentence would be true but the former false.)

Strategy E offers this answer: a 'whether'-clause *is* a disjunction. Let us introduce an operator 'wheth' which is similar to 'whether' except that it attaches to a single sentence rather than a list of two or more. It makes abnormal sentences; its semantic rule is a simplification of the rule for 'whether', as follows.

$$\mathsf{F}_{i,j} \; wheth \; A \; \text{if and only if} \; \mathsf{F}_i A \; \text{and} \; \mathsf{F}_j A.$$

Or, in terms of content, if A is normal:

If $\mathsf{F}_j A$, then $[\![wheth \; A]\!]_j = [\![A]\!]_j$;

If $\mathsf{F}_j A$, then $[\![wheth \; A]\!]_j$ is the empty set.

Then a 'whether'-clause is equivalent to a truth-functional disjunction of 'wheth'-clauses; 'whether' is 'wheth' plus 'or'. For instance:

$$\mathsf{F}_{i,j} \; whether \; A \; or \; B \; \text{if and only if} \; \mathsf{F}_{i,j} \; wheth \; A \; or$$
$$\mathsf{F}_{i,j} \; wheth \; B$$
$$\text{if and only if} \; \mathsf{F}_{i,j} \; (wheth \; A \lor wheth \; B).$$

Of course, this is a bit artificial; in that respect the original form of strategy E, without the reduction of 'whether' to 'wheth', is more attractive.

I noted that the step to strategy E (or even D) was a step in the wrong direction so far as constructions of dependence and independence are concerned. But it is not a fatally wrong step, for it is reversible should the need arise. Without treating the second family of constructions in any detail, I shall simply note that given the function that acts as the semantic value of a 'whether'-clause in strategy E – namely, the function that gives the truth value at any pair of points, or alternatively the function that gives the content relative to any one point – we can recover the alternative set needed in treating the constructions in the second family. Let W be a 'whether'-clause; then the set of alternatives presented by W is $\{ [\![W]\!]_j : j$ a point $\}$, that is,

$$\{X : \exists j \ X = \{i : \vdash_{i,j} W\}\}.$$

REFERENCES

[1] Hans Kamp, "Formal Properties of 'Now' ", *Theoria* 37 (1971) 227–273.

[2] David Lewis and Stephanie R. Lewis, review of Olson and Paul, *Contemporary Philosophy in Scandinavia, Theoria* 41 (1975) 39–60.

[3] Krister Segerberg, "Two-Dimensional Modal Logic", *Journal of Philosophical Logic* 2 (1973) 77–96.

[4] Robert Stalnaker, "Assertion", in Peter Cole, ed., *Syntax and Semantics* 9 (New York: Academic Press, 1978) 315–332.

[5] Frank Vlach, " 'Now' and 'Then' ", presented to a conference of the Australasian Association for Logic, August 1970.

[6] Lennart Åqvist, *A New Approach to the Logical Theory of Interrogatives*, Part 1: *Analysis* (Uppsala: Filosofiska Studier, 1965).

[7] Lennart Åqvist, "Scattered Topics in Interrogative Logic", in J.W. Davis, D.J. Hockney, and W.K. Wilson, eds., *Philosophical Logic* (Dordrecht, Holland: D. Reidel, 1969) 114–121.

[8] Lennart Åqvist, "Revised Foundations for Imperative-Epistemic and Interrogative Logic", *Theoria* 37 (1971) 33–73.

[9] Lennart Åqvist, "On the Analysis and Logic of Questions," in R.E. Olson and A.M. Paul, eds., *Contemporary Philosophy in Scandinavia* (Baltimore: The Johns Hopkins Press, 1972) 27–39.

[10] Lennart Åqvist, "Modal Logic with Subjunctive Conditionals and Dispositional Predicates", *Journal of Philosophical Logic* 2 (1973) 1–76.

4

Probabilities of conditionals and conditional probabilities II

Adams's thesis about indicative conditionals is that their assertability goes by the conditional subjective probability of the consequent given the antecedent, in very much the same way that assertability normally goes by the subjective probability of truth.[1] The thesis is well established; the remaining question is how it may best be explained. The nicest explanation would be that the truth conditions of indicative conditionals are such as to guarantee the equality

$$(*) \quad P(A \to C) = P(C/A) = {}^{df}P(CA)/P(A)$$

whenever $P(A)$ is positive. In a previous paper,[2] I argued that this nicest explanation cannot be right. After reviewing my previous argument, I shall here extend it in order to plug some loopholes.

I

I began with a *first triviality result,* as follows. Except in a trivial case, there is no way to interpret $\to$ uniformly[3] so that (*) holds univer-

First published in *The Philosophical Review* 95 (1986), 581–589. Reprinted with kind permission from *The Philosophical Review.*

1 See Ernest W. Adams, "The Logic of Conditionals," *Inquiry* 8 (1965), pp. 166–197; *The Logic of Conditionals* (Dordrecht: Reidel, 1975).
2 "Probabilities of Conditionals and Conditional Probabilities," *The Philosophical Review* 85 (1976), pp. 297–315; reprinted (with postscripts) in my *Philosophical Papers,* Volume II (Oxford: Oxford University Press, 1986).
3 I take it to be part of the "nicest explanation" now under attack that $\to$ is to be in-

sally, for every probability function P (as well as for every antecedent A and consequent C).

But you might say that (*) doesn't have to hold universally, throughout the class of all probability functions whatever. To explain Adams's thesis, it is good enough if (*) holds only throughout the class of *belief functions:* probability functions that represent possible systems of belief.

I agree. However I cited reasons why the change of belief that results from coming to know an item of new evidence should take place by conditionalizing on what was learned; I concluded that the class of belief functions is closed under conditionalizing; and I appealed to a *second triviality result,* as follows. Except in a trivial case, there is no way to interpret → uniformly so that (*) holds throughout a class of probability functions closed under conditionalizing.

In the previous paper, that was the end of my argument. But I now think the end came too soon. I owe you answers to two more objections.

II

You might say that not just any proposition[4] could be an item of evidence. Even granted that change of belief takes place (at least sometimes) by conditionalizing on one's total evidence, that does not mean that the class of belief functions must be closed under conditionalizing generally. It is enough that the class should be closed

terpreted uniformly, the same way in the context of one probability function as in the context of another. The proposal that → has a nonuniform interpretation, so that for each P we have a $\rightarrow_P$ that satisfies (*), but in general a different $\rightarrow_P$ for different P, is a rival hypothesis, an alternative way to explain Adams's thesis. It is unscathed by my arguments here and in the previous paper. See Robert Stalnaker, "Indicative Conditionals," *Philosophia* 5 (1975), pp. 269–286, reprinted in W. L. Harper *et al.,* eds., *Ifs* (Dordrecht: Reidel, 1981); and Bas van Fraassen, "Probabilities of Conditionals," in W. L. Harper and C. A. Hooker, eds., *Foundations of Probability Theory, Statistical Inference and Statistical Theories of Science,* Volume I (Dordrecht: Reidel, 1976).

4 Here I switch from sentences (in the previous paper) to propositions as the bearers of subjective probability. The reason is that our items of evidence might be propositions that have no accurate expression in any language we are capable of using.

under conditionalizing on those propositions that could be someone's total evidence.

In proving the second triviality result, I did not use the full strength of my assumption that the class in question was closed under conditionalizing. I only conditionalized twice: once on a proposition C, once on its negation $-C$. But that, you might well say, was bad enough. A proposition that could be someone's total evidence must be, in certain respects, highly specific. But to the extent that a proposition is specific, its negation is unspecific. So one time or the other, whether on C or on $-C$, I conditionalized on a proposition of the wrong sort; and the result may well have been a probability function that was not a belief function.

I agree. I did not give any good reason why belief functions should be closed under conditionalizing generally. I should only have assumed that they are closed under conditionalizing on a certain special class of *evidence propositions*.

Consider a limited class of evidence propositions: those that characterize the total evidence available to a particular subject at a particular time, as it is or as it might have been, in a maximally specific way. These propositions – as always with the maximally specific alternatives for a single subject matter – comprise a partition: they are mutually exclusive and jointly exhaustive. Further, the subject's limited powers of discrimination will ensure that this partition is a finite one.

So the hypothesis now before us says that (*) holds throughout the class of belief functions, and that this class is closed under conditionalizing on the members of a certain finite partition of evidence propositions. Against this I appeal to a *third triviality result*, as follows. Except in a trivial case, there is no way to interpret $\rightarrow$ uniformly so that (*) holds throughout a class of probability functions closed under conditionalizing on the propositions in some finite partition.

The proof differs little from my previous proofs of the first and second results. Let P be a probability function in the class. Let C, D, . . . be all the propositions in the partition to which P assigns positive probability; we assume that there are at least two such propositions. Let A be a proposition such that $P(A/C)$, $P(A/D)$, . . . , are all positive, and such that $P(C/A) \neq P(C)$. If there are no such P, C, D,

. . . , and *A*, that is the case I am here calling 'trivial'. We may be sure that the case of the class of belief functions and a partition of evidence propositions will not be thus trivial.

By finite additivity, the definition of conditional probability, and the incompatibility of *C*, *D*, . . . , we have that

$$P(A{\rightarrow}C) = P(A{\rightarrow}C/C)P(C) + P(A{\rightarrow}C/D)P(D) + \ldots$$

Suppose for *reductio* that (*) holds throughout the class. By applying it in turn to *P* and to all the functions $P(-/C)$, $P(-/D)$, . . . that come from *P* by conditionalizing on *D*, *C*, . . . respectively, we have

$$\frac{P(CA)}{P(A)} = \frac{P(CA/C)P(C)}{P(A/C)} + \frac{P(CA/D)P(D)}{P(A/D)} + \ldots$$

The first term of the right-hand sum simplifies and the other terms vanish, so we have

$$P(C/A) = P(C),$$

which contradicts our choice of *A* and thereby completes the *reductio*.

III

So if you grant that change of belief takes place (at least sometimes) by conditionalizing, and only insist that the conditionalizing must be on evidence propositions, then I take your point but I can make my case against (*) despite it. However you might go further. You might say that conditionalizing never yields one belief function from another.

Why? You might say it because you run my own argument in reverse.[5] Adams's thesis is compelling; it needs to be explained; it turns out that even a modicum of closure under conditionalizing would wreck the explanation; therefore even that modicum of closure is to be rejected. I reply that if (*) does not hold throughout the belief

5 Thus Wayne Davis, in discussion.

60

functions, that wrecks only one explanation of Adams's thesis. Other hypotheses remain.[6]

You might better say it for a more direct reason.[7] If one belief function came from another by conditionalizing on the evidence, it would be *irregular*. It would not spread its probability over all the possibilities there are, but would assign probability zero to all possibilities that conflict with the evidence. You might insist that an irregular probability function cannot represent a reasonable system of belief. It would be overconfident to respond to evidence by conditionalizing and thereby falling into irregularity. To do so is to dismiss the genuine possibility that one has mistaken the evidence. It is to be ready to bet against that possibility at any odds, no matter how high the stakes. And it is to dismiss the possibility forever, no matter how much later evidence may call for reconsideration. For once a proposition gets probability zero, no subsequent conditionalizing ever can restore it to positive probability. (And neither can the generalized conditionalizing we shall soon consider.) Surely it would be more reasonable to respond to evidence in a slightly more moderate way, almost conditionalizing but not quite. If a possibility conflicts with the evidence, by all means bring its probability down very low – but never quite to zero.

It's one thing to say, as you have, that an irregular probability function cannot represent a reasonable system of belief; it's something else to say that it cannot represent a system of belief at all. The latter is what you need if, despite my triviality results so far, you still say that (*) holds throughout the class of all belief functions. But you can cross the gap by appealing to a theory of intentionality according

---

6 Of several available proposals to explain Adams's thesis without the aid of (*), I now favor that of Frank Jackson, "On Assertion and Indicative Conditionals," *The Philosophical Review* 87 (1979), pp. 565–589. In particular, I take it to be better (for the reasons Jackson gives) than the explanation of Adams's thesis that I proposed in "Probabilities of Conditionals and Conditional Probabilities."

7 Thus Anthony Appiah, "The Importance of Triviality," *The Philosophical Review* 95 (1986), pp. 209–231. But Appiah is no upholder of (*), although he faults my argument against it. He rejects it on the strength of an argument advanced by I. F. Carlstrom and C. S. Hill in their review of Adams, "The Logic of Conditionals," *Philosophy of Science* 45 (1978), pp. 155–158.

to which – roughly speaking – the content of a total mental state is the system of belief and desire that best rationalizes the behavior to which that state would tend to dispose one. How could an unreasonable belief function ever be part of the system that best rationalizes anything? If it cannot, it never is suited to be part of the content of any mental state, in which case it is not a belief function after all. This theory of reason as constitutive of content must be handled with care, lest it prove too much – namely, that there is no such thing as unreason – but it may well be that a sufficiently qualified form of it would meet your needs.

(What if some state would dispose one to behavior that would indeed suggest a complete and irrevocable dismissal of the possibility that one has mistaken the evidence? Even so, it is not clear that an irregular belief function is called for. If a system is to do well at rationalizing behavior, of course one *desideratum* is that the behavior should fit the prescriptions of the system. But another *desideratum,* you may plausibly say, is that the system itself should be reasonable. Even if an irregular function does best on fit, a not-quite-irregular function may yet do better on balance.)

For what it is worth, I would insist that the ideally rational agent does conditionalize on his total evidence, and thereby falls into irregularity. He never does mistake the evidence, wherefore he may and he must dismiss the possibility that he has mistaken it. Else there is a surefire way to drain his pockets: sell him insurance against the mistakes he never makes, collect the premium, never pay any claims.[8] This surprising conclusion does nothing to show that we, the imperfect folk who do run the risk of mistaking our evidence, would be reasonable to conditionalize ourselves into irregularity. Not at all – rather, it shows that we differ more from the ideally rational agent

8 In other words, if the ideally rational agent did not conditionalize, and reserved some probability for the hypothesis that he had mistaken his evidence, then he would be vulnerable to a "diachronic Dutch book" – which, by definition of ideal rationality, he is not. See Paul Teller, "Conditionalization and Observation," *Synthese* 26 (1973), pp. 218–258, and Bas van Fraassen, "Belief and the Will," *Journal of Philosophy* 81 (1984), pp. 235–256, for discussion of diachronic Dutch books. However, Teller and van Fraassen would accept the conclusion I draw only in much-qualified form, if at all.

than might have been thought. After all, many of our virtues consist of being good at coping with our limitations. (Whatever the divine virtues may be, they can scarcely include curiosity, fortitude, or gumption.) Likewise our cautious refusal to conditionalize quite all the way helps us to cope with our fallibility. One departure from the perfection of ideal rationality demands another.

(If I am right that the ideally rational agent conditionalizes, then I have shown at least that (*) does not hold throughout the class of *his* belief functions. So if Adams's thesis applies to his indicative conditionals, it must be otherwise explained. But what does that show about us? You might well doubt whether Adams's thesis does apply to the ideally rational agent, given that we differ from him more than might have been thought.)

But if we are talking of everyday, less-than-ideal rationality, then I agree that it is unreasonable to conditionalize, reasonable to shun irregularity. Nor can I challenge your further conclusion that no belief function – anyhow, none suited to ordinary folk – is irregular. I accept it, if not wholeheartedly, at least for the sake of the argument.

If we do not conditionalize on our evidence, what do we do instead? To this question, Jeffrey has offered a highly plausible answer.[9] In conditionalizing, the probability of the evidence proposition is exogenously raised to one; the probability of anything else conditional on that evidence is left unchanged. Jeffrey generalizes this in a natural way, as follows. The probability of the evidence proposition is exogenously raised, not to one but to some less extreme value; the probability of the negation of the evidence proposition is lowered by an equal amount, as must be done to preserve additivity; and the probabilities of anything else conditional on the evidence and conditional on its negation both are left unchanged. So if P is the original probability function and the probability of proposition C is exogenously raised by amount x, we get a new probability function P_x given by the schema

$$(GC) \quad P_x(B) = P(B) + x[P(B/C) - P(B/-C)]$$

9 Richard C. Jeffrey, *The Logic of Decision* (New York: McGraw Hill, 1965, and Chicago: University of Chicago Press, 1983), Chapter 11. Jeffrey actually describes

whenever $P(C)$ and $P(-C)$ and x are positive, C is an evidence proposition, and x is less than $P(-C)$. (The final restriction excludes the case that $x = P(-C)$, which reduces to ordinary conditionalizing on C and engenders irregularity; and also the case that x is greater than $P(-C)$, in which P_x would assign negative values and so would not be a genuine probability function.) I shall assume that change of belief under the impact of evidence takes place (at least sometimes) in conformity to schema (GC).

So the hypothesis now before us says that (*) holds throughout the class of belief functions, and that this class is closed under change in conformity to schema (GC). Against this I appeal to a *fourth triviality result*, as follows. Except in a trivial case, there is no way to interpret → uniformly so that (*) holds throughout a class of probability functions closed under change in conformity to (GC).

Let P be a probability function in the class. Let C be an evidence proposition such that $P(C)$ and $P(-C)$ are positive. Let A be a proposition such that $P(A/C)$ and $P(A/-C)$ are positive and unequal. If there are no such P, C, and A, that is the case I am now calling 'trivial'. We may be sure that the class of belief functions will not be trivial in this way.

Suppose for *reductio* that (*) holds throughout the class, hence both for P and for any P_x with x between 0 and $P(-C)$. So we have

$$P(A \rightarrow C) = P(CA)/P(A),$$
$$P_x(A \rightarrow C) = P_x(CA)/P_x(A);$$

where by (GC),

$$P_x(A \rightarrow C) = P(A \rightarrow C) + x[P(A \rightarrow C/C) - P(A \rightarrow C/-C)],$$
$$P_x(CA) = P(CA) + x[P(CA/C) - P(CA/-C)]$$
$$= P(CA) + xP(A/C),$$
$$P_x(A) = P(A) + x[P(A/C) - P(A/-C)].$$

a more general case, in which there may be exogenous change to the probabilities of several evidence propositions simultaneously, but for the argument to follow it is enough to consider the simpler case. For further discussion, see Paul Teller, *op. cit.*

From these equations, we can derive that whenever x is between 0 and $P(-C)$,

$$x[P(A/C) - P(A/-C)][P(A{\rightarrow}C/C) - P(A{\rightarrow}C/-C)]$$
$$= P(A/C) - P(A)[P(A{\rightarrow}C/C) - P(A{\rightarrow}C/-C)] -$$
$$P(C/A)[P(A/C) - P(A/-C)].$$

The only way this can hold for a range of values of x is that the coefficient of x on the left-hand side is zero, and the right-hand side also is zero. There are two cases.

Case 1. The coefficient is zero because $P(A/C) = P(A/-C)$. But that contradicts our choice of A.

Case 2. The coefficient is zero because $P(A{\rightarrow}C/C) = P(A{\rightarrow} C/-C)$. Then since the right-hand side also is zero, and since its second term vanishes, we have that

$$0 = P(A/C) - P(C/A)[P(A/C) - P(A/-C)]$$
$$= P(A/C)[1 - P(C/A)] + P(C/A)\,P(A/-C)$$
$$= P(A/C)\,P(-C/A) + P(C/A)\,P(A/-C),$$

which contradicts our choice of P, C, and A, whereby all of $P(A/C)$, $P(-C/A)$, $P(C/A)$, and $P(A/-C)$ must be positive. This completes the *reductio*.

5

Intensional logics without iterative axioms

I. INTRODUCTION

Among the formulas that have been used as axioms for various intensional logics, some are *iterative:* intensional operators occur within the scopes of intensional operators. Others are *non-iterative*. Some examples from modal, deontic, and tense logics are as follows.

Iterative axioms	Non-iterative axioms
$\Diamond p \supset \Box \Diamond p$	$\Box p \supset p$
$(p \dashv q) \supset (\Box p \dashv \Box q)$	$(p \dashv q) \supset (\Box p \supset \Box q)$
$\bigcirc (\bigcirc p \supset p)$	$- (\bigcirc p \mathbin{\&} \bigcirc - p)$
$\mathsf{F}p \supset - \mathsf{P} - \mathsf{F}p$	$\mathsf{F}(p \supset p)$

It may be that there is no way to axiomatize an intensional logic without recourse to one or more iterative axioms; then I call that logic *iterative*. But many familiar logics can be axiomatized by means of non-iterative axioms alone; these logics I call *non-iterative*.

A *frame*, roughly speaking, is a partial interpretation for an intensional language. It provides a set *I*, all subsets of which are to be regarded as eligible propositional values or truth sets for formulas. It also provides, for each intensional operator in the language, a

First published in *The Journal of Philosophical Logic* 3 (1974), 457–466. Copyright © 1974 by D. Reidel Publishing Company, Dordrecht-Holland. Reprinted with kind permission from Kluwer Academic Publishers.

I am grateful to Peter Gärdenfors for valuable discussion of the subject of this paper.

function specifying how the propositional values of formulas compounded by means of that operator are to depend on the propositional values of their immediate constituents. A frame plus an assignment of values to propositional variables are enough to yield an assignment of propositional values to all formulas in the language. Iff every interpretation based on a certain frame assigns the propositional value I (truth everywhere) to a certain formula, then we say that the formula is *valid* in the frame.

There are two different natural ways in which an intensional logic L may be said to 'correspond' to a class C of frames. For any class C of frames, there is a logic L that has as its theorems exactly those formulas that are valid in all frames in the class C. We say then that the class C *determines* the logic L. Also, for any logic L there is a class C of exactly those frames that validate all theorems of L; I shall say then that C is the *range* of L.

Any class that determines a logic L must be, of course, a subclass of the range of L. A logic is usually determined by many different subclasses of its range. But note that if any class whatever determines a logic L, then in particular the range of L determines L. (If a formula is valid in all frames in the range then it is valid in all frames in the determining subclass, so it is a theorem of L.) We call a logic *complete* iff some class of frames determines it; equivalently, iff its range determines it. We know that not every logic is complete.[1] If L is incomplete, then the range of L determines not L itself but some stronger logic L'. (Then also this one class is the range of the two different logics L and L'.) One would like to know what sorts of logics

---

1 See Kit Fine, 'An Incomplete Logic Containing S4' and S. K. Thomason, 'An Incompleteness Theorem in Modal Logic', both forthcoming in *Theoria*; also M. Gerson, 'The Inadequacy of the Neighborhood Semantics for Modal Logic', unpublished. It should, however, be noted that the situation changes if we think of a frame as providing a domain of propositional values which may *not* be the full power set of any set I. Under this changed concept of a frame, all classical intensional logics are complete; see Bengt Hansson and Peter Gärdenfors, 'A Guide to Intensional Semantics', in *Modality, Morality and Other Problems of Sense and Nonsense: Essays Dedicated to Sören Halldén* (Lund, 1973). Still broader general completeness results of the same sort have been obtained by Richard Routley and Robert K. Meyer.

can be incomplete. A partial answer is: iterative ones only. *Every non-iterative logic is complete.* Somewhat more precisely: every non-iterative classical propositional intensional logic is complete with respect to classical frames with unrestricted valuations.

The proof is an adaptation of the method of filtrations. As a bonus, it yields a decidability result and an easy route to more specialized completeness results for non-iterative logics.

II. LANGUAGE

Consider a formal language partially specified as follows. (1) There are countably many propositional variables $p, q, r, \ldots$; and (2) propositional variables are formulas. (3) There are enough of the truth-functional connectives: & and $-$, let us say. (The rest may be regarded as metalinguistic abbreviations.) (4) There are at most countably many intensional operators, each of which is an n-ary connective for some number $n, n \geq 0$. (5) The result of applying any n-ary truth-functional connective or intensional operator to any n formulas (in appropriate order and with appropriate punctuation) is a formula. (6) There are no other formulas.

A formula is *iterative* iff it has a subformula of the form $0(\emptyset_1 \ldots \emptyset_n)$ where 0 is an intensional operator and at least one of the formulas $\emptyset_1 \ldots \emptyset_n$ already contains an occurrence of some intensional operator. Note that I, unlike some authors, am *not* excluding iterative formulas from the language itself. They are perfectly well-formed; and the logics we shall consider will all have iterative formulas among their theorems. It is only as axioms that iterative formulas will be forbidden.

Since there are only countably many formulas, we may suppose given some arbitrary fixed enumeration of them; thus we may legitimately refer to the *first* formula of a given sort.

III. LOGICS

We identify a logic with the set of its theorems. A set of formulas is a *logic* (more fully, a *classical propositional intensional logic*) iff it is closed under these three rules (taking 'theorem' as 'member of the set'):

TF. If ∅ follows by truth-functional logic from zero or more theorems, then ∅ is a theorem.

Substitution. If ∅ is the result of uniformly substituting formulas for propositional variables in a theorem, then ∅ is a theorem.

Interchange. If $\psi \equiv \chi$ is a theorem, and if ∅' is the result of substituting ψ for χ at one or more places in ∅, then $\emptyset \equiv \emptyset'$ is a theorem.

A set of formulas A is an *axiom set for* a logic L iff L is the least logic that includes all members of A as theorems. A logic L is *iterative* iff every axiom set for L contains at least one iterative formula; L is *non-iterative* iff at least one axiom set for L contains no iterative formula. A logic L is *finitely axiomatizable* iff some axiom set for L is finite.

IV. FRAMES, INTERPRETATIONS, VALIDITY

A *frame* (or more fully, a *classical frame with unrestricted valuations*) is any pair $\langle I, f \rangle$ of a set I and a family f of functions, indexed by the intensional operators of the language, such that if 0 is an n-ary operator then f_0 is a function from n-tuples of subsets of I to subsets of I.

An *interpretation* is any function $[\![\,]\!]$ which assigns to each formula ∅ of the language a set $[\![\emptyset]\!]$. (We may think of $[\![\emptyset]\!]$ as the truth set or propositional value of ∅ under the interpretation.) An interpretation $[\![\,]\!]$ is *based on* a frame $\langle I, f \rangle$ iff the following conditions hold for all formulas and operators:

(0) $[\![\emptyset]\!] \subseteq I,$
(1) $[\![\emptyset \,\&\, \psi]\!] = [\![\emptyset]\!] \cap [\![\psi]\!],$
(2) $[\![-\emptyset]\!] = I - [\![\emptyset]\!],$
(3) $[\![0(\emptyset_1 \ldots \emptyset_n)]\!] = f_0([\![\emptyset_1]\!] \ldots [\![\emptyset_n]\!]).$

A formula ∅ is *valid in* a frame $\langle I, f \rangle$ iff, for every interpretation $[\![\,]\!]$ based on $\langle I, f \rangle$, $[\![\emptyset]\!] = I$. If C is a class of frames and L is a logic (in the precise senses we have now defined), we recall that C *determines* L iff the theorems of L are all and only the formulas valid in all frames in the class C; and that C is the *range* of L iff C contains all and only those frames in which all theorems of L are valid. (We can also verify at this point that every class C of frames determines a logic; that is, that the set of formulas valid in all frames in C is closed under *TF, Substitution,* and *Interchange.*)

We shall need the following two lemmas; I omit the obvious proofs.

LEMMA 1. Suppose that A is an axiom set for a logic L, and that all formulas in A are valid in a frame $\langle I,f \rangle$. Then all theorems of L are valid in $\langle I,f \rangle$ (which is to say that $\langle I,f \rangle$ belongs to the range of L).

LEMMA 2. Suppose that $\emptyset'$ is the result of substituting formulas ψ_p, ψ_q, ... uniformly for the propositional variables p, q, ... in a formula $\emptyset$. Suppose that $[\![\ ]\!]$ and $[\![\ ]\!]'$ are two interpretations based on the same frame such that $[\![\psi_p]\!] = [\![p]\!]'$, $[\![\psi_q]\!] = [\![q]\!]'$, and so on. Then $[\![\emptyset']\!] = [\![\emptyset]\!]'$.

V. FILTRATIONS

Let L be any logic. A set of sentences is *L-consistent* iff no theorem of L is the negation of any conjunction of sentences in the set. A *maximal L-consistent* set is an L-consistent set that is not properly included in any other one. For any logic L, we have these familiar lemmas.

LEMMA 3. Any L-consistent set is included in a maximal L-consistent set.

LEMMA 4. If i is any maximal L-consistent set, then (1) a conjunction $\emptyset \ \& \ \psi$ belongs to i iff both conjuncts $\emptyset$ and ψ belong to i; (2) a negation $-\emptyset$ belongs to i iff $\emptyset$ does not belong to i; (3) a biconditional $\emptyset \equiv \psi$ belongs to i iff both or neither of $\emptyset$ and ψ belong to i.

LEMMA 5. Any theorem of L belongs to all maximal L-consistent sets.

Next, let θ be any formula. A *θ-description* is any set of zero or more subformulas of θ together with the negations of all other subformulas of θ. For every L-consistent θ-description D, let i_D be an arbitrarily chosen one of the maximal L-consistent supersets of D; and let I be the set of all these i_D's. I contains one member for each L-consistent θ-description. But there are only 2^s θ-descriptions in all, where s is the number of subformulas of θ. Wherefore we have:

LEMMA 6. I is finite, and an upper bound on its size can be computed by examination of θ.

A theorem of L must belong to all maximal L-consistent sets, and hence to all members of I. We have a restricted converse:

LEMMA 7. Suppose that $\emptyset$ is a truth-functional compound of subformulas of θ. Then $\emptyset$ belongs to all members of I iff $\emptyset$ is a theorem of L.

Proof: If D is a θ-description and $\emptyset$ is a truth-functional compound of subformulas of θ, then $D \cup \{\emptyset\}$ and $D \cup \{-\emptyset\}$ cannot both be consistent in truth-functional logic, and so cannot both be L-consistent. *Case 1*: there is some θ-description D such that $D \cup \{-\emptyset\}$ is L-consistent. Then $\emptyset$ is not a theorem of L. Then also $D \cup \{\emptyset\}$ is not L-consistent, and hence $\emptyset$ does not belong to i_D. *Case 2*: there is no θ-description D such that $D \cup \{-\emptyset\}$ is L-consistent. Then $\emptyset$ is a theorem of L, by the rule *TF*. Then also $\emptyset$ must belong to every i_D in I. Q.E.D.

It will be convenient to introduce the following notation: let $/\emptyset/$ $= df\{i \in I : \emptyset \in i\}$.

We next specify an assignment of formulas, called *labels,* to all members and subsets of I. Any member of I is i_D for some one L-consistent θ-description $D;$ let the label of i_D be the first conjunction (first in our arbitrary enumeration of all formulas) of all formulas in D. Any non-empty subset of I contains finitely many members; let the label of the subset be the first disjunction of the labels of all of its members. Finally, let the label of the empty set be the contradiction $\theta \& -\theta$. Our assignment of labels has the following properties.

LEMMA 8. Every subset of I has as its label a truth-functional compound of subformulas of θ.

LEMMA 9. If $\emptyset$ is the label of a subset J of I, then $/\emptyset/ = J$.

We next specify a family f of functions such that $\langle I, f \rangle$ is a frame. Let 0 be any n-ary operator, let $J_1 \ldots J_n$ be any n subsets of I, and let $\emptyset_1 \ldots \emptyset_n$ be the labels of $J_1 \ldots J_n$ respectively; then $f_0(J_1 \ldots J_n)$ is to be $/0(\emptyset_1 \ldots \emptyset_n)/$.

Next, let $[\![\ ]\!]$ be the interpretation based on the frame $\langle I, f \rangle$ such that the following equation (*) holds whenever the formula ø is a propositional variable:

(*) $[\![ø]\!] = /ø/$.

We call the pair of the specified frame $\langle i, f \rangle$ and the specified interpretation $[\![\ ]\!]$ based on that frame a *filtration* (*for* the logic L and the formula θ).

Filtrations are of interest because the equation (*) carries over from propositional variables to certain compound formulas, in accordance with the following two lemmas.

LEMMA 10. (*) holds for any truth-functional compound of constituent formulas for which (*) holds.

The proof is immediate. We need only consider conjunctions and negations, and compare clauses (1) and (2) of the definition of an interpretation based on a frame with parts (1) and (2) of Lemma 4. We turn next to compounding by means of the intensional operators.

LEMMA 11. (*) holds for a formula of the form $0(ø_1 \ldots ø_n)$ whenever (*a*) (*) holds for each of the constituent formulas $ø_1 \ldots ø_n$, and (*b*) $ø_1 \ldots ø_n$ are all truth-functional compounds of subformulas of θ.

Proof: Let $\psi_1 \ldots \psi_n$ be the labels of the sets $[\![ø_1]\!] \ldots [\![ø_n]\!]$ respectively. By definition of f, $[\![0(ø_1 \ldots ø_n)]\!] = f_0([\![ø_1]\!] \ldots [\![ø_n]\!]) = /0(\psi_1 \ldots \psi_n)/$. Also, by Lemma 9 and hypothesis (*a*), $/\psi_1/ = /ø_1/, \ldots,$ and $/\psi_n/ = /ø_n/$. Consider the biconditionals $\psi_1 \equiv ø_1, \ldots, \psi_n \equiv ø_n$. By part (3) of Lemma 4, they belong to all members of I; by hypothesis (*b*) and Lemma 8 they are truth-functional compounds of subformulas of θ; so it follows by Lemma 7 that they are theorems of L. By repeated use of Interchange, $0(\psi_1 \ldots \psi_n) \equiv 0(ø_1 \ldots ø_n)$ also is a theorem of L; wherefore by Lemma 7 and part (3) of Lemma 4, $/0(\psi_1 \ldots \psi_n)/ = /0(ø_1 \ldots ø_n)/$. Q.E.D.

The corollaries which one usually aims at in working with filtrations are these:

LEMMA 12. (*) holds for all subformulas of θ.

LEMMA 13. (*) holds for θ itself.

But our lemmas on truth-functional and intensional compounding overshoot the subformulas of θ, and this overshoot is crucial for our present task. We have this further corollary:

LEMMA 14. Let $\emptyset$ be a non-iterative formula, let $\psi_p, \ldots \psi_q, \ldots$ be truth-functional compounds of subformulas of θ, and let $\emptyset'$ be the result of substituting $\psi_p, \psi_{q'} \ldots$ uniformly for the propositional variables $p, q, \ldots$ in $\emptyset$. Then (*) holds for $\emptyset'$.

Lemma 14 will apply, in particular, when $\emptyset$ is a non-iterative axiom and $\psi_p, \psi_q, \ldots$ are labels of various subsets of I.

LEMMA 15. Suppose that $\emptyset$ is a non-iterative formula, and that $\emptyset$ belongs to some axiom set for the logic L. Then $\emptyset$ is valid in $\langle I, f \rangle$.

Proof: We must show that for every interpretation $[\![\ ]\!]'$ based on $\langle I, f \rangle$, $[\![\emptyset]\!]' = I$. Given any such $[\![\ ]\!]'$, let $\emptyset'$ be the result of substituting the label ψ_p of the set $[\![p]\!]'$ uniformly for the propositional variable p, the label ψ_q of $[\![q]\!]'$ uniformly for $q, \ldots$, in $\emptyset$. Since (*) holds for each of these labels, we have $[\![\psi_p]\!] = /\psi_p/ = [\![p]\!]'$, $[\![\psi_q]\!] = /\psi_q/ = [\![q]\!]'$, and so on; so by Lemma 2, $[\![\emptyset]\!]' = [\![\emptyset']\!]$. By Lemma 14, $[\![\emptyset']\!] = /\emptyset'/$. Since $\emptyset$ is an axiom and hence a theorem of L, it follows by Substitution that $\emptyset'$ also is a theorem of L, so $/\emptyset/ = I$. Q.E.D.

LEMMA 16. If L is a non-iterative logic, then $\langle I, f \rangle$ is in the range of L.

Proof: Let A be an axiom set for L that contains no iterative formulas. By Lemma 15, all formulas in A are valid in $\langle I, f \rangle$, so by Lemma 1 all theorems of L are valid in $\langle I, f \rangle$. Q.E.D.

LEMMA 17. If θ is not a theorem of L, θ is invalid in $\langle I, f \rangle$.

Proof: If θ is not a theorem of L, then by Lemmas 7 and 13 $[\![\theta]\!] \neq I$. Q.E.D.

We can summarize our lemmas on filtrations as follows.

THEOREM 1. If L is any non-iterative logic and θ is any non-theorem of L, then there is a frame $\langle I, f \rangle$, belonging to the range of L, in which θ is invalid. Further, I is finite and an upper bound on its size can be computed by examination of θ.

VI. GENERAL COMPLETENESS AND DECIDABILITY OF NON-ITERATIVE LOGICS

Our general completeness theorem follows at once.

THEOREM 2. Every non-iterative logic is complete.

As for decidability, it suffices to note that if a logic L is finitely axiomatizable, then it is a decidable question whether or not a given formula θ is valid in all frames $\langle I, f \rangle$ in the range of L such that the size of I does not exceed a bound computed by examination of θ. Although there are infinitely many such frames (except in trivial cases), we may ignore irrelevant differences among them: differences between isomorphic frames, or between frames that differ only with respect to operators that do not appear in θ. It is enough to search through finitely many representative cases, all of them decidable. Therefore we have:

THEOREM 3. Every finitely axiomatizable non-iterative logic is decidable.

VII. SPECIAL COMPLETENESS RESULTS

We often wish to establish not only that a logic is complete but also that it is determined by some subclass of its range that is of special interest to us. If we have to do with non-iterative logics, Theorem 1 facilitates the proof of special completeness results by allowing us to ignore problems that arise only in infinite cases. To show that a class C determines a non-iterative logic L, it is enough to show that C is intermediate in inclusiveness between the full range of L and the part of the range consisting of finite frames; in other words, to show (1) that C is a subclass of the range of L, and (2) that $\langle I, f \rangle$ belongs to C whenever $\langle I, f \rangle$ is in the range of L and I is finite.

By way of example, consider the non-iterative regular modal logics. Let our language contain a single, unary operator $\Box$; and let L be a logic having as a theorem

C. $\Box(p \ \& \ q) \equiv (\Box p \ \& \ \Box q)$.

Any such logic is called a *regular modal logic*. Further, let L be non-iterative. The frames of special interest to us will be those that derive from the standard accessibility semantics for modal logic. A *modal structure* is a triple $\langle I, N, R \rangle$ of a set I, a subset N of I (the *normal* set), and a relation R (the *accessibility* relation) included in $N \times I$. A frame $\langle I, f \rangle$ *corresponds* to such a modal structure iff, for any subset A of I, $f_\Box A = \{i \in N:$ whenever $iRj, j \in A\}$. That gives the standard semantic rule: if an interpretation $[\![\ ]\!]$ is based on a frame corresponding to a modal structure, then always i belongs to $[\![\Box\emptyset]\!]$ iff i is normal and whenever j is accessible from i, j belongs to $[\![\emptyset]\!]$. The correspondence thus defined is one-one between all modal structures and exactly those frames $\langle I, f \rangle$ that obey the following distributive condition: whenever $\mathscr{A}$ is a set of subsets of I, then $f_\Box(\cap \mathscr{A}) = \cap\{f_\Box A : A \in \mathscr{A}\}$. (Let $\langle I, f \rangle$ be such a frame; then the corresponding modal structure is obtained by taking $N = f_\Box I$, iRj iff i belongs to N and whenever i belongs to $f_\Box A, j$ belongs to A.) Not every frame in the range of L satisfies this distributive condition; to validate **C,** it is enough that the distributive condition should hold for finite $\mathscr{A}$. But when the frame itself is finite, the case of infinite $\mathscr{A}$ does not arise. Thus if $\langle I, f \rangle$ is in the range of L and I is finite, then $\langle I, f \rangle$ satisfies the distributive condition and corresponds to a modal structure. The subclass of the range of our non-iterative regular modal logic L comprising those frames that correspond to modal structures is therefore a determining class for L. From here we can go instantly to the standard completeness results for such logics as **C2, D2, E2, K, D,** and **T.**

For another example, consider the non-iterative ones of the V-logics discussed in my *Counterfactuals* (Blackwell, 1973). Let our language contain a single, binary operator $\leqslant$; and let L be a logic having as theorems

Trans. $(p \leqslant q \ \& \ q \leqslant r) \supset p \leqslant r,$

Connex. $p \leqslant q \vee q \leqslant p,$

Dis. $((p \vee q) \leqslant r) \equiv (p \leqslant r \vee q \leqslant r).$

I call any such logic a *V-logic*. (My Rule for Comparative Possibility in *Counterfactuals* is interderivable with the combination used here of **Dis** and the rule of Interchange.) Further, let L be non-iterative. The frames $\langle I, f \rangle$ of special interest to us will be those derived from families of rankings of I. A *ranking structure* $\langle K_i, R_i \rangle_{i \in I}$ is an assignment to each i in I of a subset K_i of I (the *evaluable* set) and a weak ordering R_i of K_i (the *ranking*). The frame $\langle I, f \rangle$ *corresponds* to such a ranking structure iff, for any subsets A and B of I, $f_{\leqslant}(A, B) = \{i \in I:$ for every k in $B \cap K_i$ there is some j in $A \cap K_i$ such that $jR_i k\}$. That gives the semantic rule: if an interpretation $[\![\ ]\!]$ is based on a frame corresponding to a ranking structure, then always i belongs to $[\![\emptyset \leqslant \psi]\!]$ iff, relative to i, for every evaluable k in $[\![\psi]\!]$, some evaluable j in $[\![\emptyset]\!]$ is ranked at least as highly as k is. A frame corresponds to every ranking structure, but not conversely; however, if $\langle I, f \rangle$ is a frame in the range of L and I is finite, then $\langle I, f \rangle$ corresponds to the ranking structure $\langle K_i, R_i \rangle_{i \in I}$ obtained as follows. For any i in I, let K_i be $\{j \in I: i \notin f_{\leqslant}(\wedge, \{j\})\}$; for any i in I and j, k in K_i, let $jR_i k$ iff i belongs to $f_{\leqslant}(\{j\}, \{k\})$. The subclass of the range of our non-iterative V-logic L comprising those frames that correspond to ranking structures is therefore a determining class for L. From here we can instantly reach the completeness results given in *Counterfactuals* for those V-logics that can be axiomatized without the non-iterative axioms **U** and **A**.

6

Ordering semantics and premise semantics for counterfactuals

1. COUNTERFACTUALS AND FACTUAL BACKGROUND

Consider the counterfactual conditional 'If I were to look in my pocket for a penny, I would find one'. Is it true? That depends on the factual background against which it is evaluated. Perhaps I have a penny in my pocket. Its presence is then part of the factual background. So among the possible worlds where I look for a penny, those where there is no penny may be ignored as gratuitously unlike the actual world. (So may those where there is only a hidden penny; in fact my pocket is uncluttered and affords no hiding place. So may those where I'm unable to find a penny that's there and unhidden.) Factual background carries over into the hypothetical situation, so long as there is nothing to keep it out. So in this case the counterfactual is true. But perhaps I have no penny. In that case, the absence of a penny is part of the factual background that carries over into the hypothetical situation, so the counterfactual is false.

Any formal analysis giving truth conditions for counterfactuals must somehow take account of the impact of factual background. Two very natural devices to serve that purpose are orderings of

First published in *The Journal of Philosophical Logic* 10 (1981), 217–234. Copyright © 1981 by D. Reidel Publishing Company, Dordrecht, Holland, and Boston, U.S.A. Reprinted with kind permission from Kluwer Academic Publishers.

I thank Max Cresswell, Angelika Kratzer, and Robert Stalnaker for valuable discussions. I thank Victoria University of Wellington and the New Zealand–United States Educational Foundation for research support.

worlds and sets of premises. Ordering semantics for counterfactuals is presented, in various versions, in Stalnaker [8], Lewis [5], and Pollock [7]. (In this paper, I shall not discuss Pollock's other writings on counterfactuals.) Premise semantics is presented in Kratzer [3] and [4]. (The formally parallel theory of Veltman [11] is not meant as truth-conditional semantics, and hence falls outside the scope of this discussion.) I shall show that premise semantics is equivalent to the most general version – roughly, Pollock's version – of ordering semantics.

I should like to postpone consideration of the complications and disputes that arise because the possible worlds are infinite in number. Let us therefore pretend, until further notice, that there are only finitely many worlds. That pretence will permit simple and intuitive formulations of the theories under consideration.

2. ORDERING SEMANTICS IN THE FINITE CASE

We may think of factual background as ordering the possible worlds. Given the facts that obtain at a world i, and given the attitudes and understandings and features of context that make some of these facts count for more than others, we can say that some worlds fit the facts of world i better than others do. Some worlds differ less from i, are closer to i, than others. Perhaps we do not always get a clear decision between two worlds that depart from i in quite different directions; the ordering may be only partial. And perhaps some worlds differ from i so much that we should ignore them and leave them out of the ordering altogether. But sometimes, at least, we can sensibly say that world j differs less from i than world k does.

In particular, i itself differs not at all from i, and clearly no world differs from i less than that. It is reasonable to assume, further, that any other world differs more from i than i itself does.

Given a counterfactual to be evaluated as true or false at a world, such an ordering serves to divide the worlds where the antecedent holds into two classes. There are those that differ minimally from the given world; and there are those that differ more-than-minimally, gratuitously. Then we may ignore the latter, and call the counterfactual true iff the consequent holds throughout the worlds in the former class.

The ordering that gives the factual background depends on the facts about the world, known or unknown; how it depends on them is determined – or underdetermined – by our linguistic practice and by context. We may separate the contribution of practice and context from the contribution of the world, evaluating counterfactuals as true or false at a world, and according to a 'frame' determined somehow by practice and context.

We define an *ordering frame* as a function that assigns to any world i a strict partial ordering $\angle_i$ of a set S_i of worlds, satisfying the condition:

(Centering) i belongs to S_i; and for any j in S_i, $i \angle_i j$ unless $i = j$.

(*A strict partial ordering* of a set is a transitive, asymmetric, binary relation having that set as its field.) We call j a *closest A-world to i* (according to an ordering frame) iff (i) j is an A-world, that is, a world where proposition A holds, (ii) j belongs to S_i, and (iii) there is no A-world k such that $k \angle_i j$. We can lay down the following truth condition for a counterfactual from A to C (that is, one with an antecedent and consequent that express the propositions A and C, respectively): the counterfactual is true at world i, according to an ordering frame, iff

(OF) C holds at every closest A-world to i.

Truth condition OF is common to all versions of ordering semantics, so long as we stick to the finite case; what I have said so far is neutral between the theories of Stalnaker, Lewis, and Pollock.

The three theories differ in various ways. Some of the differences concern informal matters: is it correct or misleading to describe the orderings that govern the truth of counterfactuals as comparing the 'overall similarity' of worlds? Just how is the appropriate ordering determined by features of the compared worlds, and by practice and context? How much residual indeterminacy is there? These questions fall mostly outside the scope of this paper. Other differences concern extra formal requirements that might be imposed on ordering frames. So far, following Pollock, we have allowed merely partial orderings, in which worlds may be tied or incomparable. Lewis permits ties but prohibits incomparabilities; Stalnaker prohibits ties and incomparabilities both. We shall consider these disagreements in Section 5. Further differences concern the postponed difficulties of the infinite case, and we shall consider these in Section 6.

The restricted field S_i of the ordering $\angle_i$ might well be a needless complication. We could get rid of it by imposing an extra condition on ordering frames:

(Universality) the field S_i of $\angle_i$ is always the set of all worlds.

I know of no very strong reasons for or against imposing the requirement of Universality, and accordingly I shall treat it as an optional extra. Suppose we are given an ordering frame that does not satisfy Universality, and suppose we would prefer one that does. The natural adjustment is as follows. Where the original frame assigns to i the ordering $\angle_i$ of S_i, let the new frame assign the ordering $\angle_i^+$ of all worlds, where $j\angle_i^+ k$ iff either $j\angle_i k$ or j does and k does not belong to S_i. Call the new frame the *universalisation* of the original frame. The difference between the two frames only matters when the antecedent of a counterfactual is true at some worlds, but not at any of the worlds in S_i. Then the original frame treats the counterfactual as vacuous, making it true at i regardless of its consequent; whereas the universalisation makes it true at i iff the consequent is true at every world where the antecedent is, so that the antecedent strictly implies the consequent.

Some further notation will prove useful. Let us write $\langle \angle_i \rangle$ for the ordering frame that assigns to any world i the ordering $\angle_i$ (S_i can be left out, since it is the field of $\angle_i$). Let us also write $j\leqq_i k$ to mean that either $j\angle_i k$ or j is identical to k and belongs to S_i. The relation $\leqq_i$ is then a nonstrict partial ordering of S_i. And let us write $\sim_i$ to mean that neither $j\angle_i k$ nor $k\angle_i j$ although both j and k belong to S_i. If so, j and k are either identical, or tied as differing equally from i, or incomparable with respect to their difference from i. The relation $\sim_i$ is reflexive and symmetric, but not in general transitive.

A simpler way to think of factual background is as a set of facts – a set of true propositions about the world, which together serve to distinguish it from all other worlds. These facts serve as auxiliary premises which may join with the antecedent of a counterfactual to imply the consequent, thereby making the counterfactual true against that factual background. The obvious problem is that some of the facts will contradict the antecedent (unless it is true), so the entire set of them will join with the antecedent to imply anything whatever. We must therefore use subsets of the factual premises, cut down just enough to be consistent with the antecedent. But how shall we cut the premise set down – what goes, what stays? We might invoke some system of weights or priorities, telling us that some of the factual premises are to be given up more readily than others. That would lead directly back to ordering semantics. Alternatively, we might treat the premises equally. We might require that the cut-down premise set must join with the antecedent to imply the consequent no matter how the cutting-down is done; all ways must work. That is the approach to counterfactuals taken by Kratzer in [4]; it is a special case of her treatment of conditional modality in [3], which in turn is based on her treatment of modality in [2].

We must be selective in the choice of premises. If we take all facts as premises, then (as is shown in [4]) we get no basis for discrimination among the worlds where an antecedent holds and the resulting truth condition for counterfactuals is plainly wrong. By judicious selection, we can accomplish the same sort of discrimination as would result from unequal treatment of premises. As Kratzer explains in [4], the outcome depends on the way we lump items of information together in single premises or divide them between several premises. Lumped items stand or fall together, divided items can be given up one at a time. Hence if an item is lumped into several premises, that makes it comparatively hard to give up; whereas if it is confined to a premise of its own, it can be given up without effect on anything else. This lumping and dividing turns out to be surprisingly powerful as a method for discriminating among worlds – so much so that, as will be shown, premise semantics can do anything that ordering semantics can. Formally, there is nothing to choose. Intuitively, the question is whether the same premises that it would seem natural to select are the ones that lump and divide properly; on that question I shall venture no opinion.

Let us identify a proposition with the set of worlds where it holds. Logical terminology applied to propositions has its natural set-theoretic meaning: conjunctions and disjunctions are intersections and unions, the necessary and impossible propositions are the set of all worlds and the empty set, consistency among propositions is nonempty intersection, implication is inclusion of the intersection of the premises in the conclusion, and so on. Facts about a world i are sets of worlds that contain i, and a set of facts rich enough to distinguish i from all other worlds is one whose intersection has i as its sole member.

Again we distinguish the contribution of the world from the contribution of a 'frame' determined somehow by linguistic practice and context. The world provides the facts, the frame selects some of those facts as premises. A counterfactual is evaluated at a world, and according to a frame.

We define a *premise frame* $\langle H_i \rangle$ as a function that assigns to any world i a set H_i of propositions – *premises* for i – satisfying the condition:

(Centering) i does, and all other worlds do not, belong to every proposition in H_i.

An *A-consistent premise set* for i is a subset of H_i that is consistent with the proposition A; and it is a *maximal A-consistent premise set* for i iff, in addition, it is not properly included in any larger A-consistent premise set for i. We can lay down the following truth condition for a counterfactual from A to C: it is true at world i, according to a premise frame, iff

(PF)   whenever J is a nonempty maximal A-consistent premise set for i, J and A jointly imply C.

82

This is almost, but not quite, Kratzer's truth condition for the finite case; hers is obtained by deleting the word 'nonempty'. The reason for the change is explained below.

Worlds where none of the premises in H_i hold are ignored in evaluating counterfactuals at i, just as worlds outside S_i are in ordering semantics. If the antecedent of a counterfactual holds only at such ignored worlds, it is just as if the antecedent holds at no worlds at all: there are no nonempty maximal A-consistent premise sets, so the counterfactual is true regardless of its consequent. If we find it objectionable that some worlds are thus ignored, we could do as Kratzer does and delete the word 'nonempty', so that the counterfactual will be true at i iff A (jointly with the empty set) implies C. Alternatively, we could impose an extra condition on premise frames to stop the problem from arising:

(Universality)   every world belongs to some proposition in H_i.

I shall treat this requirement of Universality as an optional extra, like the corresponding requirement of Universality in ordering semantics.

Suppose we have a premise frame $\langle H_i \rangle$ that does not satisfy Universality, and we want one that does; then we can simply take a new frame $\langle H_i^+ \rangle$ that assigns to any world i the set consisting of all the propositions in H_i and the necessary proposition as well. Call $\langle H_i^+ \rangle$ the *universalisation* of $\langle H_i \rangle$. Except in the case considered, in which the antecedent holds only at some worlds outside all the premises for i, a premise frame and its universalisation evaluate all counterfactuals alike. In effect, Kratzer opts for Universality; but she builds it into her truth condition instead of imposing it as an extra requirement on premise frames. We are free to start with a frame that does not satisfy Universality, but we universalise it whenever we use it: a counterfactual is true at a world according to $\langle H_i \rangle$ under Kratzer's truth condition iff it is true at that world according to $\langle H_i^+ \rangle$ under the neutral truth condition PF. I think it better to use PF, both for the separation of distinct questions and for easy comparison with extant versions of ordering semantics. If we want to consider exactly Kratzer's version of premise semantics, we need only impose Universality as a requirement on frames.

83

Given a premise frame $\langle H_i \rangle$, there is a natural way to derive from it an ordering frame $\langle \angle_i \rangle$: let S_i be the union of the propositions in H_i; and for any j and k in S_i, let $j \angle_i k$ iff all propositions in H_i that hold at k hold also at j, but some hold at j that do not hold also at k. The worlds that can be ordered are those where at least some of the premises hold; a closer world conforms to all the premises that a less close world conforms to and more besides. If each $\angle_i$ is derived in this way from the corresponding H_i, it is easily seen that $\langle \angle_i \rangle$ must be an ordering frame. Let us call the frames $\langle H_i \rangle$ and $\langle \angle_i \rangle$ *equivalent*.

Equivalent frames evaluate counterfactuals alike, at least in the finite case. Let $\langle H_i \rangle$ and $\langle \angle_i \rangle$ be equivalent frames. Then for any propositions A and C and any world i, PF holds iff OF holds.

> *Proof.* Let j be any A-world and let J be the set of all propositions in H_i that hold at j. It is enough to show that (i) J is a nonempty maximal A-consistent premise set for i iff (ii) j is a closest A-world to i. We may assume that J is nonempty, else (i) and (ii) are both false. ($\rightarrow$) If not (ii), we have $k \in A \cap S_i$ such that $k \angle_i j$. Let K be the set of propositions in H_i that hold at k. K is an A-consistent premise set for i and it properly includes J, so not (i). ($\leftarrow$) If not (i), J must be properly included in some larger A-consistent premise set for i, call it K. Take any k in $A \cap \cap K$. Then $k \angle_i j$, so not (ii). Q.E.D.

By definition, every premise frame is equivalent to some unique ordering frame. However, two premise frames may be equivalent to the same ordering frame. Suppose two premise frames are alike except that one assigns to i the premises $\{i\}$ and $\{i,j,k\}$ while the second assigns to i the premises $\{i,j\}$ and $\{i,k\}$. Either way, the derived ordering $\angle_i$ is the same: S_i is $\{i,j,k\}$, $i \angle_i j$, $i \angle_i k$, and $j \sim_i k$. This means that premise frames contain surplus information – information that makes no difference to the way the premise frames do their job of evaluating counterfactuals. Intuitively, this surplus information concerns the difference between ties and incomparabilities. The first of our frames represents j and k as alike in the way they differ from i, whereas the second represents them as departing from i in different directions. Ordering frames – if they use strict orderings, as in my present formulation – omit this surplus information. Any two of them disagree in their evaluation of some counterfactual at some world, and from the fact that $j \sim_i k$ we cannot tell whether to regard j and k as tied or as incomparable.

Every ordering frame can be derived from an equivalent premise frame. Thus the equivalence of premise and ordering frames is a many-one correspondence, exhausting both classes.

> *Proof.* Suppose given an ordering frame $\langle \angle_i \rangle$. For each world i, let H_i be $\{\{j : j \leq_i k\} : k \in S_i\}$. Centering for $\langle H_i \rangle$ follows from Centering for $\langle \angle_i \rangle$: any i belongs to each member of H_i, and $\{i\}$ belongs to H_i since it is $\{j : j \leq_i i\}$, so $\cap H_i$ is $\{i\}$. Hence $\langle H_i \rangle$ is a premise frame. Further, $\langle H_i \rangle$ and $\langle \angle_i \rangle$ are equivalent. Each S_i is $\cup H_i$: for any $k \in S_i$ we have $k \in \{j : j \leq_i k\} \in H_i$, and for any $j \in \cup H_i$ we have $j \in S_i$ since $j \leq_i k \in S_i$. Also, for any g and h in S_i, $g \angle_i h$ iff g belongs to all the members of H_i that h belongs to and more besides. ($\rightarrow$) If h belongs to $\{j : j \leq_i k\}$, then so does g since $g \angle_i h \leq_i k$. But g belongs to $\{j : j \leq_i g\}$ and h does not, and $g \in S_i$. ($\leftarrow$) Since $h \in S_i$ and h belongs to $\{j : j \leq_i h\}$, so does g. But $g \neq h$ since they do not belong to exactly the same sets, so $g \angle_i h$. Q.E.D.

Finally, it is immediate from the definition of equivalence that if $\langle H_i \rangle$ and $\langle \angle_i \rangle$ are equivalent frames, then the former satisfies the optional Universality requirement of premise semantics iff the latter satisfies the optional Universality requirement of ordering semantics. Further, any two equivalent frames have equivalent universalisations.

5. PARTIAL VERSUS MULTIPLE ORDERINGS

The results of Section 4 show that premise semantics is equivalent to a version of ordering semantics – Pollock's version, if we ignore questions that arise in the infinite case – in which the orderings may be merely partial. Kratzer joins Pollock, and seemingly disagrees with Stalnaker and Lewis, in permitting worlds to differ from a given world in incomparable ways. However, I shall argue that there is less to this disagreement than meets the eye.

Although an ordering frame cannot, and need not, distinguish incomparabilities from ties, it can nevertheless reveal that incomparabilities are present. Recall that the relation $\sim_i$ is reflexive and symmetric. Unless it is transitive as well, some cases in which $j \sim_i k$ must be incomparabilities, not ties or identities; for the relation of being tied or identical must surely be an equivalence relation. If $\sim_i$ is transitive, on the other hand, there is no bar to regarding it as the relation of being tied or identical. Let us do so. To prohibit incomparabilities, as Lewis and Stalnaker do, is to impose the following as an extra requirement on ordering frames:

(Comparability) $\sim_i$ is transitive.

That means that each $\angle_i$ must be at least a strict weak ordering. (Lewis's formulation, in [5]: 48–50, actually uses nonstrict weak orderings, but it is equivalent to the formulation considered here.) To prohibit ties as well, as Stalnaker does, is to impose the stronger requirement that $\sim_i$ is just the relation of identity among worlds in S_i, or equivalently:

(Trichotomy) for any j and k in S_i, $j \angle_i k$ or $j = k$ or $k \angle_i j$.

That means that each $\angle_i$ is a strict simple ordering.

Pollock [7] argues that no ordering frame without incomparabilities could give the intuitively correct evaluations of all the counterfactuals in a certain small set. But Pollock's argument is suspect, as noted by Loewer [6]: 111. Pollock's example involves English counterfactual sentences with seemingly disjunctive antecedents; such sentences behave oddly in a way that would account for Pollock's evidence with no need for any incomparabilities.

A better reason to question Comparability, however, is close at hand. Ordinary counterfactuals usually require only the comparison of worlds with a great deal in common, from the standpoint of worlds of the sort we think we might inhabit. An ordering frame that satisfies Comparability would be cluttered up with comparisons that matter to the evaluation of counterfactuals only in peculiar cases that will never arise. Whatever system of general principles we use to make the wanted comparisons will doubtless go on willy nilly to make some of the unwanted comparisons as well, but it seems not likely that it will settle them all (not given that it makes the wanted comparisons in a way that fits our counterfactual judgements). An ordering frame that satisfies Comparability would be a cumbersome thing to keep in mind, or to establish by our linguistic practice. Why should we have one? How could we? Most likely we don't.

This argument is persuasive, and I know of no one who would dispute it. However, it is not exactly an argument against Comparability. Rather, it is an argument against the combination of Comparability with full determinacy of the ordering frame. We need no partial orderings if we are prepared to admit that we have not bothered to decide quite which total orderings (weak or simple, as the case may be) are the right ones. The advocates of Comparability certainly are prepared to admit that the ordering frame is left underdetermined by linguistic practice and context. (Stalnaker [9]; Lewis [5]: 91–95.) So where Pollock sees a determinate partial ordering that leaves two worlds incomparable, Stalnaker and Lewis see a multiplicity of total orderings that disagree in their comparisons of the two worlds, with nothing in practice and context to select one of these orderings over the rest. Practice and context determine a class of frames each satisfying Comparability, not a single frame that fails to satisfy Comparability.

Formally, there is nothing to choose, so long as we are concerned only with conditions of determinate truth for counterfactuals, and so long as we stick to the finite case. Ordering semantics with and without Comparability, or even Trichotomy, are in a sense equivalent. Premise semantics is in the same sense equivalent not only to Pollock's version of ordering semantics but also to Lewis's, and even to Stalnaker's.

Compare the difference between Lewis and Stalnaker: where Lewis sees a tie, Stalnaker sees indeterminacy between simple orderings that break that tie in opposite ways. A counterfactual is true on Lewis's semantics iff it is true on Stalnaker's semantics no matter how the ties are broken. (See Lewis [5]: 81–83; and for a fuller discussion, van Fraassen [10].) The same method of reconciliation applies also to the seeming disagreement about Comparability. (We shall consider ordering semantics only, but of course what follows carries over to premise semantics by equivalence of frames.) Where Pollock or Kratzer sees an incomparability, Lewis or Stalnaker sees indeterminacy between total (weak or simple) orderings that make the missing comparison in all possible ways. A counterfactual is true on Pollock's or Kratzer's semantics iff it is true on Lewis's or Stalnaker's semantics no matter how the missing comparisons are made.

Let us call frame $\langle \angle_i^* \rangle$ a *refinement* (alternatively, a *Stalnaker refinement*) of frame $\langle \angle_i \rangle$ iff (i) for any i, S_i^* is the same as S_i, (ii) whenever $j \angle_i k$, then $j \angle_i^* k$, and (iii) $\langle \angle_i^* \rangle$ satisfies Comparability (alternatively, Trichotomy). We wish to show that a counterfactual is true at a world according to $\langle \angle_i \rangle$ iff it is true at that world according to every refinement (alternatively, every Stalnaker refinement) of $\langle \angle_i \rangle$, on truth condition OF. (Here I assume that the antecedent and consequent will express the same proposition according to the original frame and the refinements; that may not be so if they are themselves counterfactuals, or compounded in part from counterfactuals.)

> *Proof.* It suffices to show that (i) j is a closest A-world to h according to some refinement (alternatively, some Stalnaker refinement) $\langle \angle_i^* \rangle$ of $\langle \angle_i \rangle$ iff (ii) j is a closest A-world to h according to $\langle \angle_i \rangle$. Assume $j \in A \cap S_h$, else (i) and (ii) are both false. ($\rightarrow$) If not (ii), we have $k \in A$ such that $k \angle_h j$. Then also $k \angle_h^* j$ for any refinement (alternatively, any Stalnaker refinement) $\langle \angle_i^* \rangle$ of $\langle \angle_i \rangle$, so not (i). ($\leftarrow$) We will construct a Stalnaker refinement $\langle \angle_i^* \rangle$ of $\langle \angle_i \rangle$ such that whenever $j \sim_h k$ and $j \neq k$, then $j \angle_h^* k$. Suppose for *reductio* that j is not a closest A-world to h according to $\langle \angle_i^* \rangle$. Then we have $k \in A$ such that $k \angle_h^* j$. Not $k \angle_h j$ by (ii). Not $j \angle_h k$ since then $j \angle_h^* k \angle_h^* j$. Not $j = k$ since then $k \angle_h^* k$. Not both $j \sim_h k$ and $j \neq k$ since then again $j \angle_h^* k \angle_h^* j$. But there is no other alternative, so the supposition is refuted. It remains to construct the refinement $\langle \angle_i^* \rangle$. If $i \neq h$, let $\angle_i^0 = \angle_i$. Let $f \angle_h^0 g$ iff either $f \angle_h g$ or there is some k such that $j \sim_h k$, $j \neq k$, $f \leq_h j$, and $k \leq_h g$. Now for each i, take an arbitrary sequence $\langle f_0, g_0 \rangle$, $\langle f_1, g_1 \rangle$, ... of all pairs of worlds in S_i, and form a parallel sequence $\angle_i^0$, $\angle_i^1$, ... as follows. We have $\angle_i^0$. If $f_n \angle_i^n g_n$ or if $f_n = g_n$ or if $g_n \angle_i^n f_n$, let $\angle_i^{n+1} = \angle_i^n$. Otherwise, let $f \angle_i^{n+1} g$ iff either $f \angle_i^n g$ or both $f \leq_i^n f_n$ and $g_n \sim_h^n g$. Let $\angle_i^*$ be the last term of this sequence. (In the infinite case, invoke the axiom of choice to take the pairs in an arbitrary transfinite sequence; and in forming the parallel sequence, take unions at limit ordinals and at the end.) Transitivity and asymmetry are preserved at each step, and in the preliminary step from $\angle_h$ to $\angle_h^0$ (and in taking unions, in the infinite case), so each $\angle_i^*$ is (at least)

88

a strict partial ordering. At each step the field of the ordering remains S_i. Also $\angle_i$ is included in $\angle_i^*$, since we only add pairs and never remove any. Hence $\langle \angle_i^* \rangle$ satisfies Centering, and so is a frame. Also any pair $\langle f, g \rangle$ will be added at some step unless $\langle g, f \rangle$ is present first or $f = g$, so $\langle \angle_i^* \rangle$ satisfies Trichotomy and is a Stalnaker refinement of $\langle \angle_i \rangle$. Finally, whenever $j \sim_h k$ and $j \neq k$, $j \angle_h^{\shortparallel} k$ and hence $j \angle_h^* k$. Q.E.D.

Although there is no remaining issue about conditions of determinate truth for counterfactuals in the finite case – determinate truth being truth according to all frames within the range of indeterminacy – some lesser questions remain in dispute. What are we to say of a counterfactual that is false according to the original frame, true according to some of its refinements, and false according to other refinements? Is it false or is it indeterminate in truth value? What are we to say of its negation? What of a disjunction of counterfactuals such that the original frame makes neither disjunct true, but every refinement makes one or the other disjunct true? I think these questions are best answered by a version of ordering semantics that prohibits all incomparabilities but permits at least some ties, however I shall not argue that case here.

6. THE INFINITE CASE

Our definitions and results all carry over to the infinite case. However, the adequacy of the truth conditions OF and PF becomes doubtful. For all we have said so far, there might be infinite sequences without limit: in ordering semantics, ever-closer A-worlds to i but no closest one, or in premise semantics, ever-larger A-consistent premise sets for i but no maximal one. Then OF and PF give the absurd result that any counterfactual whose antecedent expresses A is true at i regardless of its consequent.

There is a choice between two remedies, and on this question we change partners. Lewis and Kratzer prefer to modify the truth condition, so that we get reasonable evaluations even in these troublesome cases. We shall consider shortly how this may be done. Stalnaker and Pollock prefer not to modify the truth condition, but rather to make sure that the troublesome cases never arise. They impose an extra requirement on ordering frames:

(Limit Assumption) unless no A-world belongs to S_i, there is some closest A-world to i.

We could likewise impose an extra requirement on premise frames:

(Limit Assumption) unless A is consistent with no proposition in H_i, there is some nonempty maximal A-consistent premise set for i.

It follows immediately from our first result in Section 4 that these two formulations agree: A at i violates the Limit Assumption for a premise frame iff A at i violates the Limit Assumption for the equivalent ordering frame. Hence one of two equivalent frames satisfies the Limit Assumption iff the other does.

The Limit Assumption makes infinite frames no more troublesome than finite ones. It makes it safe to continue to use our original truth conditions OF and PF, which certainly are simpler and more intuitive than the modified versions of Lewis and Kratzer. It has the further advantage, regarded as decisive by Pollock [7] and Hertzberger [1], of validating certain plausible principles of the infinitary logic of counterfactuals:

(Consistency Principle) if C_1, C_2, . . . are inconsistent, then so are the counterfactuals from A to C_1, C_2, . . . (except that they may be vacuously true together, in which case all counterfactuals from A are true);

(Consequence Principle) if C_1, C_2, . . . jointly imply B, then the counterfactuals from A to C_1, C_2, . . . jointly imply the counterfactual from A to B.

These principles can fail under the modified truth conditions of Lewis and Kratzer when the set of C_n's is infinite and there is a violation of the Limit Assumption.

Despite these formal advantages, I still think it best not to impose the Limit Assumption. The trouble with it is that it is apt to conflict with any account we might wish to give of how the orderings or the premise sets that comprise a frame are determined. Conflicts can arise in a variety of ways. The example of the line, given in [5]: 20–21, shows how the Limit Assumption may fail if we base our orderings on a weighted sum of degrees of similarity or dissimilarity in various respects. Pollock [7] draws the conclusion that we should not construct the orderings that way – he puts his point by saying that they are not similarity orderings – but I think he underestimates the problem. The Limit Assumption can fail in quite different ways also. It can fail even if we stick to atomistic, all-or-nothing respect of similarity and difference, and even if we give up Comparability rather than balancing these off against one another.

> *Example.* Consider a premise frame that assigns to world i an infinite set H_i of independent propositions, so that any conjunction of some of these propositions with the negations of all the rest is consistent. Consider also the equivalent ordering frame, according to which $j \angle_i k$ iff j conforms to all the premises in H_i that k conforms to and more besides. Thus $\angle_i$ is rife with incomparabilities. These frames fail to satisfy the Limit Assumption. Let A hold at exactly those worlds where infinitely many propositions in H_i fail to hold. A premise set for i is A-consistent iff it leaves out infinitely many members of H_i, and there is no maximal such set. Likewise there is no closest A-world to i.

If we want the Limit Assumption, I take it that what we need is some sort of coarse-graining. We must imitate the finite case by ignoring most of the countless respects of difference that make the possible worlds infinite in number. Coarse-graining is certainly a formal option; whether it can be built into an intuitively adequate analysis of counterfactuals seems to me to be an open question. Therefore I think it best to remain neutral on the Limit Assumption, and to replace truth conditions OF and PF by modified versions that do not need the Limit Assumption to work properly.

Lewis's version of ordering semantics uses a truth condition, given in [5]: 49, which does not need the Limit Assumption: a counterfactual from A to C is true at world i, according to an ordering frame, iff

(OC)   unless no A-world belongs to S_i, there is some A-world
j in S_i such that for any A-world k in S_i, either C holds
at k or $j \angle_i k$.

But OC is not satisfactory for our present purposes. Although it does not need the Limit Assumption, it does need Comparability. It might misevaluate a counterfactual as false – say, one that is true under OF in the finite case – because the worlds where the antecedent holds divide into incomparable classes.

We want a fully neutral truth condition: one that needs neither the Limit Assumption nor Comparability. I propose the following: a counterfactual from A to C is true at world i, according to an ordering frame, iff

(O)   for any A-world h in S_i, there is some A-world j such that
(i) $j \leqq_i h$, and (ii) C holds at any A-world k such that
$k \leqq_i j$.

This condition is fully neutral. However, it is a generalization both of OF and of OC, reducing to the former given the Limit Assumption and to the latter given Comparability.

Proof. Assume that $A \cap S_i$ is nonempty, else O, OF, and OC all are true. (1) O implies OF. Let h be any A-world closest to i. By O we have $j \in A$ such that (i) $j \leqq_i h$, and hence $j = h$, and (ii) C holds at any A-world k such that $k \leqq_i j$, and hence at h itself. (2) OF and the Limit Assumption imply O. Given $h \in A \cap S_i$, let B be the set of all worlds $g \in A$ such that $g \leqq_i h$. By the Limit Assumption, since $h \in B \cap S_i$, we have j which is a closest B-world to i. Since $j \in B$, $j \leqq_i h$. Also j is a closest A-world to i, for if $f \in A$ and $f \angle_i j$, then $f \in B$ and so j is not a closest B-world. So $j \in C$ by OF. Whenever $k \in A$ and $k \leqq_i j$, it must be that $k = j$ and hence $k \in C$. (3) O and Comparability imply OC. By O, since $A \cap S_i$ is nonempty, we have some $j \in A \cap S_i$ such that whenever $k \in A$ and $k \leqq_i j$ then $k \in C$. If for any $k \in A \cap S_i$ either $k \in C$ or $j \angle_i k$, we are done. If not, we have $g \in A \cap S_i$ such that neither $g \in C$ nor $j \angle_i g$. Not $g \leqq_i j$, else $g \in C$, so $g \sim_i j$. By O again, we have $j' \in A$ such that (i) $j' \leqq_i g$, and (ii) whenever $k \in A$ and $k \leqq_i j'$ then $k \in C$, so in particular $j' \in C$. Then $j' \neq g$ so $j' \angle_i g$. By Comparability, $j' \angle_i j$. For any $k \in A \cap S_i$, either

$j' \angle_i k$, or $k \sim_i j'$ in which case by Comparability $k \angle_i j$ and so $k \in C$, or $k \angle_i j'$ in which case again $k \angle_i j$ and so $k \in C$. (4) OC implies O. By OC we have $g \in A \cap S_i$ such that for any $k \in A \cap S_i$ either $k \in C$ or $g \angle_i k$. For any $h \in A \cap S_i$, let $j = g$ if $g \leq_i h, j = h$ otherwise. Either way, (i) $j \leq_i h$, and (ii) whenever $k \in A$ and $k \leq_i j$, not $g \angle_i k$ so $k \in C$. Q.E.D.

As a neutral truth condition for premise semantics, we can take the following: a counterfactual from A to C is true at i, according to a premise frame, iff

(P)   For any nonempty A-consistent premise set H for i, there is some A-consistent premise set J for i such that (i) H is included in J, and (ii) J and A jointly imply C.

This is almost, but not quite, Kratzer's truth condition in [4] for the infinite case. As in the finite case, hers is obtained by deleting the word 'nonempty'; it has Universality built in, so that a counterfactual is true at a world according to a frame $\langle H_i \rangle$ under Kratzer's truth condition iff it is true at that world according to the universalization $\langle H_i^+ \rangle$ under truth condition P.

When we switch to the neutral truth conditions O and P, it remains true that equivalent frames evaluate counterfactuals alike. (From that and previous results, it follows that P reduces to PF given the Limit Assumption, as we would wish.)

Let $\langle H_i \rangle$ and $\langle \angle_i \rangle$ be equivalent frames. Then for any propositions A and C and any world i, P holds iff O holds.

Proof. Assume that there is some nonempty A-consistent premise set for i, and hence some world in $A \cap S_i$; else P and O both are false. ($\rightarrow$) Let h be any world in $A \cap S_i$, and let H be the set of propositions in H_i that hold at h. H is a nonempty A-consistent premise set for i. By P we have an A-consistent premise set J for i such that (i) $H \subseteq J$, and (ii) J and A jointly imply C. If $J = H$ let $j = h$, otherwise let j be any world in $A \cap \cap J$. Either way, $j \in A$ and $j \leq_i h$. Whenever $k \in A$ and $k \leq_i j$, $k \in \cap J$ so $k \in C$. ($\leftarrow$) Let H be any nonempty A-consistent premise set for i, and let h be any world in $A \cap \cap H$. Then $h \in A \cap S_i$, so by O we have $j \in A$ such that

93

(i) $j \leqq_i h$, and (ii) whenever $k \in A$ and $k \leqq_i j$ then $k \in C$. Let J be the set of propositions in H_i that hold at j. Whether $j = h$ or $j \angle_i h$, $H \subseteq J$. J is an A-consistent premise set for i. So if J and A jointly imply C, we are done. If not, there is some world $h' \in A \cap \cap J$ such that $h' \notin C$. Since $h' \in A \cap S_i$, by O we have $j' \in A$ such that (i) $j' \leqq_i h'$, and (ii) whenever $k \in A$ and $k \leqq_i j'$ then $k \in C$, so in particular $j' \in C$ and hence $j' \neq h'$. Then $j' \angle_i h'$. Let J' be the set of propositions in H_i that hold at j'; $H \subseteq J \subseteq J'$ and $J \neq J'$. J' is an A-consistent premise set for i. Consider any $k \in A \cap \cap J'$; $k \angle_{i} j$, so $k \in C$. Hence J' and A jointly imply C. Q.E.D.

Unfortunately, one thing we lose in switching to our modified truth conditions is the reconciliation about Comparability that we considered in Section 5. Under O, we can have a counterfactual true at a world according to a frame, but not according to all refinements of that frame.

Example. Let the A-worlds in S_i be indexed by the integers: . . . , $j_{-2}, j_{-1}, j_0, j_1, j_2, ...$ Let $j_m \angle_i j_n$ iff m is even and $n = m + 1$; let $j_m \angle_i^* j_n$ iff $m \angle n$ in the usual ordering of the integers; and let C hold at j_m iff m is even. Then under O the counterfactual from A to C is true at i according to a frame that assigns $\angle_i$ to i, but not according to a refinement that assigns $\angle_i^*$.

Perhaps some more complicated version of the reconciliation would succeed, but I fear the complication would make it at least somewhat *ad hoc*.

7. FRAMES WITHOUT CENTERING

We could remove the Centering requirements from our definitions of ordering and premise frames, thereby redefining them in a more general sense. Generalized ordering frames are discussed in Lewis [5]: 97–121, and generalized premise frames in Kratzer [2] and [3].

By dropping Centering, we can extend ordering or premise semantics to certain relatives of counterfactuals, notably deontic conditionals. These too are evaluated against a background that may be expressed by orderings or premise sets. However, the background is deontic rather than factual: we may say that (from the standpoint of a certain world, and according to a frame somehow determined by our linguistic practice and by context) one world is better than another, or a certain premise is something that ought to hold. Lewis discusses ordering semantics for deontic conditionals in [5]: 96–104, and Kratzer's premise semantics for conditional modality in [3] applies to deontic conditionals as well as to counterfactuals. Formally, except for Centering, the ordering or premise frames and truth conditions given for deontic conditionals are like those for counterfactuals. Similar questions can arise about Universality, Comparability and Trichotomy, and the Limit Assumption versus modified truth conditions for the infinite case. Though we certainly do not want Centering, we might – but we might not – want some of its consequences, such as a requirement that all premise sets be consistent.

Our equivalence results did not depend on Centering, so they carry over to generalized ordering and premise frames. Thus premise semantics for deontic conditionals is again equivalent to a version of ordering semantics that permits merely partial orderings, and again partial orderings can in the finite case be replaced by classes of weak or simple orderings.

In fact, we made almost no use of Centering in our proofs. None, except in checking that certain structures constructed from frames that satisfied Centering were themselves frames satisfying Centering. We can now restate those results: (i) if a generalized premise frame and a generalized ordering frame are equivalent, both or neither satisfies Centering; (ii) if a generalized ordering frame is a refinement of a frame that satisfies Centering, then it too satisfies Centering.

Even in the case of ordering and premise semantics for counterfactuals, Centering might be questioned; see Lewis [5]: 28. It is clear that we must not have worlds closer to i than i itself, or premises for i that do not hold at i. It is less clear that we must not have other worlds that differ not at all from i, or where all the premises for i hold. Though I am on the whole persuaded to require Centering – in company with Stalnaker, Pollock, and Kratzer – it is worth noting the possibility of versions of ordering and premise semantics for counterfactuals in which Centering is weakened.

BIBLIOGRAPHY

[1] Hans G. Hertzberger, 'Counterfactuals and Consistency', *Journal of Philosophy* **76** (1979), 83–88.

[2] Angelika Kratzer, 'What "Must" and "Can" Must and Can Mean', *Linguistics and Philosophy* **1** (1977), 337–355.

[3] Angelika Kratzer, 'Conditional Necessity and Possibility', in R. Bäuerle, U. Egli, and A. von Stechow (eds.), *Semantics from Different Points of View* (Springer-Verlag, 1979).

[4] Angelika Kratzer, 'Partition and Revision: The Semantics of Counterfactuals', *Journal of Philosophical Logic* **10** (1981), 201–216.

[5] David Lewis, *Counterfactuals* (Blackwell, 1973).

[6] Barry M. Loewer, 'Cotenability and Counterfactual Logics', *Journal of Philosophical Logic* **8** (1979), 99–115.

[7] John L. Pollock, 'The "Possible Worlds" Analysis of Counterfactuals', *Philosophical Studies* **29** (1976), 469–476; this paper is incorporated in Pollock, *Subjunctive Reasoning* (Reidel, 1976), pp. 13–23 and 42–44.

[8] Robert Stalnaker, 'A Theory of Conditionals', in N. Rescher (ed.), *Studies in Logical Theory* (Blackwell, 1968).

[9] Robert Stalnaker, 'A Defense of Conditional Excluded Middle', in W. L. Harper, R. Stalnaker, and G. Pearce (eds.), *Ifs* (Reidel, 1980).

[10] Bas C. van Fraassen, 'Hidden Variables in Conditional Logic', *Theoria* **40** (1974), 176–190.

[11] Frank Veltman, 'Prejudices, Presuppositions and the Theory of Conditionals', in J. Groenendijk and M. Stokhof (eds.), *Amsterdam Papers in Formal Grammar*, Vol. I (Centrale Interfaculteit, Universiteit van Amsterdam, 1976).

7

Logic for equivocators

At first the champions of relevance in logic taught that relevance and preservation of truth were two separate merits, equally required of any relation that claims the name of implication. Thus Anderson and Belnap, in [1]: 19–20, cheerfully agree that certain classical implications necessarily preserve truth, but fault them for committing fallacies of relevance. (And there is no hint that necessary truth-preservation might not be truth-preservation enough.) Lately, however, relevance has been praised not – or not only – as a separate merit, but rather as something needed to ensure preservation of truth. The trouble with fallacies of relevance, it turns out, is that they can take us from truth to error.

Classical implication does preserve truth, to be sure, so long as sentences divide neatly into those that are true and those with true negations. But when the going gets tough, and we encounter true sentences whose negations also are true, then the relevant logician gets going. Then his relevant implication preserves truth and some classical implication doesn't. Say that A is true and its negation $\sim A$ is true as well. Then A & $\sim A$ is a conjunction of truths, and hence true.

First published in *Noûs* 16 (1982), 431–441. Reprinted with kind permission from Blackwell Publishers.
I thank John Burgess, Max Cresswell and J. Michael Dunn for helpful discussions.

Let B be false, and not true. Then the classically valid, but irrelevant, implication *ex falso quodlibet*

$$\frac{A \ \& \ {\sim}A}{\therefore B} \qquad \text{or} \qquad \frac{\begin{array}{c} A \\ {\sim}A \end{array}}{\therefore B}$$

fails to preserve truth. Likewise $A \lor B$ is true, being a disjunction with a true disjunct. So disjunctive syllogism, the relevantist's bugbear,

$$\frac{\begin{array}{c} A \lor B \\ {\sim}A \end{array}}{\therefore B}$$

also fails to preserve truth in this case. Whether or not we love relevance for its own sake, surely we all agree that any decent implication must preserve truth. So should we not all join the relevantists in rejecting *ex falso quodlibet* and disjunctive syllogism as fallacious? So say Belnap [2], Dunn [3] and [4], Priest [10], Routley [12] and [13], and Routley and Routley [14]. See also Makinson [7], Chapter 2, who explains this motivation for relevance without himself endorsing it.

The proposed vindication of relevance can be spelled out with all due formality. But first we must settle a verbal question. If there are truths with true negations, what is it to be false? Shall we call a sentence false if and only if it has a true negation? Or if and only if it isn't true? (In the terminology of Meyer [8]: 3–6, shall we use "American" or "Australian" valuations? In the terminology of Routley [13]: 329, note 11, shall we speak of "systemic" or "metalogical" falsity? Or rather – since we can speak of both and needn't choose – which shall get the short name "falsity"?) Like the majority of our authors, I shall speak in the first way. A truth with a true negation is false as well as true. Its negation is true (as well as false) because negation reverses truth values. We shall have a three-valued semantics; or perhaps four-valued, if we allow truth-value gaps. Two of the values are regarded as two sorts of truth: a sentence is true if and only if it is either true only or both true and false. Likewise we have two sorts of falsity: some false sentences are false only, some are both true and false. Perhaps also some sentences are neither true nor false.

Let our language be built up from atomic sentences using just the standard connectives ~, &, and ∨, governed by these orthodox rules of truth and falsity.

(For ~)   ~A is true iff A is false, false iff A is true.

(For &) A & B is true iff both A and B are true,
          false iff either A or B is false.

(For ∨) A ∨ B is true iff either A or B is true,
          false iff both A and B are false.

No classical logician could call these rules mistaken, though some might call them longwinded. But they must be longwinded if they are to apply when sentences are both true and false, or when they are neither. We may check that according to these rules, as we expected, *ex falso quodlibet* and disjunctive syllogism fail to preserve truth when A is both true and false and B is false only. Their premises are, *inter alia,* true, their conclusions are not. The story continues in three versions.

Version E (Dunn [3]; Belnap [2]; and Makinson [7]: 30–8 on "De Morgan implication"). We allow truth-value gaps, and thus we have a four-valued semantics. A premise set implies a conclusion if and only if any valuation that makes each premise true also makes the conclusion true. The implications so validated turn out to be just those given by the first-degree fragment of the Anderson-Belnap logic E of entailment. Along with *ex falso quodlibet* and disjunctive syllogism, the alleged fallacies of relevance

$$\text{(*)} \quad \frac{A}{\therefore B \vee \sim B} \qquad \text{(**)} \quad \frac{A \ \& \ \sim A}{\therefore B \vee \sim B}$$

also are not validated. They fail in case A is both true and false and B is neither.

Version RM (Dunn [4]; Dunn [3]: 166–7, note 7; and Makinson [7]: 38–9 on "Kalman implication"). We prohibit truth-value gaps and require valuations to make every sentence true or false or both; thus we have a three-valued semantics. We complicate the definition of implication, requiring not only truth but also truth-only to be preserved: (i) any valuation that makes each premise true also makes the conclusion true, and (ii) any valuation that makes each premise true and not false also makes the conclusion true and not false. The implications so validated are those given by the first-degree fragment of the partly relevant logic R-mingle. The irrelevant (**) is validated; but *ex falso quodlibet,* disjunctive syllogism, and (*) are not. Note that (*) fails for a new reason: if *A* is true only and *B* is both true and false, then (*) preserves truth but fails to preserve truth-only.

Version LP (Priest [10]). Again we have only three values: true only, false only, both true and false. But we return to the simple definition of implication, no longer requiring preservation of truth-only. As in Version E, a valid implication is one such that any valuation that makes each premise true also makes the conclusion true. The implications thus validated are given by the first-degree fragment of Priest's "logic of paradox" LP. Both (*) and (**) are validated, so the vindication of relevance on this version is very partial indeed; but the principal offenders, *ex falso quodlibet* and disjunctive syllogism, still fail. We pay dearly for our simplified definition of truth-preserving implication, since we lose contraposition. Although *A* implies $B \lor \sim B$ (irrelevantly), $\sim(B \lor \sim B)$ does not contrapositively imply $\sim A$: for a counterexample, let *B* be both true and false and let *A* be true only. Thus Priest's LP is weaker than E when it comes to contraposition, but stronger even than R-mingle when it comes to tolerating some alleged fallacies of relevance.

II

All this may help illuminate the technicalities of relevant logic; but of course it is worthless as an intuitive vindication of relevance unless somehow it makes sense that sentences can be both true and false. How can that be? Two answers are on the market. I am persuaded by neither of them.

Routley [12] and [13] and Priest [10] offer a radical answer. Maybe no special explanation is needed, but only liberation from traditional dogma. Maybe some truths just *do* have true negations. Maybe some sentences just *are* both true and false. (For Routley:

"systemically" false.) Routley and Priest mention various candidates, and others come to mind as well. (i) We have our familiar budget of logical paradoxes: the liar, the Russell set, and so on. We have very persuasive arguments that the paradoxical sentences are true, and that they are false. Why not give in, stop struggling, and simply assent to both conclusions? (ii) On some subjects, the truths and falsehoods are of our own making. The Department can make it true or false by declaration that a dissertation is accepted. If by mistake or mischief both declarations were made, might both succeed? (iii) Thorny problems of physics might prove more tractable if we felt free to entertain inconsistent solutions – the particle has position p, it has momentum q, but it does not have both position p and momentum q. (iv) Likewise for thorny problems of theology. Indeed, this proposal amounts to a generalization to other realms of the plea that "God is not bound by human logic".

This radical answer leads *prima facie* to a three-valued semantics, Version RM or Version LP. To be sure, it could be combined with provision for truth-value gaps – and intuitive motivations for gaps are a dime a dozen – to yield Version E. But it does not seem to provide a single, uniform motivation for the four-valued semantics. Even if somehow there are indeterminacies in the world as well as contradictions, there is no reason to think that the two phenomena are two sides of the same coin.

The reason we should reject this proposal is simple. No truth does have, and no truth could have, a true negation. Nothing is, and nothing could be, literally both true and false. This we know for certain, and *a priori*, and without any exception for especially perplexing subject matters. The radical case for relevance should be dismissed just because the hypothesis it requires us to entertain is inconsistent.

That may seem dogmatic. And it is: I am affirming the very thesis that Routley and Priest have called into question and – contrary to the rules of debate – I decline to defend it. Further, I concede that it is indefensible against their challenge. They have called so much into question that I have no foothold on undisputed ground. So much the worse for the demand that philosophers always must be ready to defend their theses under the rules of debate.

There is another, more conservative, answer to the question how a sentence can be both true and false. It is suggested by Dunn [3]; it appears alongside the radical answer in Routley [12]; and some of its ingredients appear in Belnap [2]. (Belnap discusses the problem of quarantining inconsistencies in the data bank of a question-answering computer, but declines to suggest that we and the computer well might solve our similar problems in similar ways.) This answer runs as follows. Never mind "ontological" truth and falsity, that is, truth and falsity *simpliciter.* Instead, consider truth and falsity according to some corpus of information. It might be someone's system of beliefs, a data bank or almanac or encyclopedia or textbook, a theory or a system of mythology, or even a work of fiction. The information in the corpus may not all be correct, and misinformation may render the corpus inconsistent. Then we might well say that sentences about matters on which there is an inconsistency are both true according to the corpus and false according to the corpus; and we might say the same of their negations. We want a conception of truth according to a corpus such that, if the corpus is mostly correct, then truth according to it will serve as a good, if fallible, guide to truth *simpliciter.* We can reasonably ask that such a conception satisfy the following desiderata. (1) Anything that is explicitly affirmed in the corpus is true according to it. (2) Truth according to the corpus is not limited to what is explicitly there, but is to some extent closed under implication. (This may include implication with the aid of background information, but let us ignore this complication.) (3) Nevertheless, an inconsistency does not – or does not always – make everything true according to the corpus. Hence truth according to the corpus is closed not under classical implication generally, but under some sort of restricted implication capable of quarantining inconsistencies. Maybe not all inconsistencies can be quarantined successfully, but many can be. (4) A sentence is false according to the corpus if and only if its negation is true according to the corpus.

This proposal looks as if it could give us an intuitive, uniform motivation for the four-valued semantics of Version E and for the resulting case against irrelevant implication. (Presumably it would not work to motivate the three-valued versions. A corpus is even more likely to be incomplete than inconsistent, so it is inevitable that some sentences will be neither true nor false according to it. To avoid such gaps, we would have to pass from real-life corpora to their completions.)

But in fact, I do not think the proposal succeeds. I agree with much of it. If a corpus is inconsistent, I see nothing wrong with saying that a sentence and its negation both may be true according to that corpus; or that such a sentence is true according to it and also false according to it. I further agree that we have a legitimate and useful conception of truth according to a corpus that satisfies the four desiderata I listed. So far, so good. But the conception I have in mind does not work in a way that fits the four-valued semantics. Nor does it use restrictions of relevance to quarantine inconsistencies. Instead, it uses a method of fragmentation. I shall explain this informally; for the technicalities, see Jaśkowski [6], Rescher and Brandom [11], and Schotch and Jennings [15]. (However, these authors' intended applications of fragmentation differ to some extent from mine.)

I speak from experience as the repository of a mildly inconsistent corpus. I used to think that Nassau Street ran roughly east-west; that the railroad nearby ran roughly north-south; and that the two were roughly parallel. (By "roughly" I mean "to within 20°".) So each sentence in an inconsistent triple was true according to my beliefs, but not everything was true according to my beliefs. Now, what about the blatantly inconsistent conjunction of the three sentences? I say that it was not true according to my beliefs. My system of beliefs was broken into (overlapping) fragments. Different fragments came into action in different situations, and the whole system of beliefs never manifested itself all at once. The first and second sentences in the inconsistent triple belonged to – were true according to – different fragments; the third belonged to both. The inconsistent conjunction of all three did not belong to, was in no way implied by, and was not true according to, any one fragment. That is why it was not true according to my system of beliefs taken as a whole. Once the fragmentation was healed, straightway my beliefs changed: now I think that Nassau Street and the railroad both run roughly northeast-southwest.

I think the same goes for other corpora in which inconsistencies are successfully quarantined. The corpus is fragmented. Something about the way it is stored, or something about the way it is used, keeps it from appearing all at once. It appears now as one consistent corpus, now as another. The disagreements between the fragments that appear are the inconsistencies of the corpus taken as a whole. We avoid trouble with such inconsistencies (and similar trouble with errors that do not destroy consistency) by not reasoning from mixtures of fragments. Something is true according to the corpus if and only if it is true according to some one fragment thereof. So we have no guarantee that implication preserves truth according to the corpus, unless all the premises come from a single fragment. What follows from two or more premises drawn from disagreeing fragments may be true according to no fragment, hence not true according to the corpus.

The details of the implication do not matter. Two-premise "fallacies of relevance" such as disjunctive syllogism, or the version of *ex falso quodlibet* in which A and $\sim A$ are separate premises, may indeed fail to preserve truth according to an inconsistent corpus. But so may many-premise implications of impeccable relevance, such as the implication from conjuncts to their conjunction (as in our example). Only one-premise implications can be trusted not to mix fragments. Irrelevant one-premise implications – such "fallacies" as (*), (**), and the one-premise version of *ex falso quodlibet* – can no more mix fragments than any other one-premise implications can. In short, to the extent that inconsistencies are quarantined by fragmentation, restrictions of relevance have nothing to do with it.

If we take truth values according to an inconsistent corpus, it may well happen that A and $\sim A$ are both true (as well as false) and B is false only. But that is not enough to give us a counterexample against *ex falso quodlibet* in its one-premise version. Such a counterexample requires truth according to the corpus of the inconsistent conjunction A & $\sim A$, not just of its conjuncts. But the conjunction need not be true according to the corpus. The corpus may be fragmented, and the conjuncts may be true according to different fragments. As we have already seen, conjunction need not preserve truth according to the corpus.

The four-valued semantics of Version E gave us orthodox rules of truth and falsity for the connectives. (Likewise for the three-valued versions.) These rules are unexceptionable for truth and falsity *simpliciter*. But if we try to understand the four-valued semantics in terms of truth and falsity according to a corpus, and if the corpus is fragmented, then the rule for & may fail. It is such a failure that wrecks our counterexample against one-premise *ex falso quodlibet*. The rule for ∨ also may fail, wrecking our previous counterexamples against (*) and (**); I spare you the details. Therefore this strategy does not lead to a successful interpretation of the four-valued semantics, even though it does provide a way in which sentences can be regarded as both true and false.

I am inclined to think that when we are forced to tolerate inconsistencies in our beliefs, theories, stories, etc., we quarantine the inconsistencies entirely by fragmentation and not at all by restrictions of relevance. In other words, truth according to any single fragment is closed under unrestricted classical implication. This view would take some strenuous defending, but its defense is irrelevant to my present purpose. For if the quarantine works only partly by fragmentation, that is enough to make my point. That is enough to make the rules for & and ∨ fail, so that the four-valued semantics of Version E will not apply. (Nor will the three-valued semantics.) Further, even if a mixed quarantine disallows the alleged fallacies of relevance, we still get no vindication of the relevant logic E. It cannot be trusted to preserve truth according to a fragmented corpus, nor can any logic that ever lets us mix fragments in many-premise implications.

A concession. I need not quarrel with anyone who wishes to put forward a fifth desideratum for a conception of truth according to a corpus: (5) the orthodox rules for & and ∨ must apply without exception. I claim no monopoly on behalf of the conception I have been discussing. I gladly agree that there are legitimate and intuitive conceptions that do satisfy desideratum (5), along with some of the original four. One such conception is obtained by closing under unrestricted classical implication; that satisfies all the desiderata except (3). Another is obtained by taking the intersection rather than the union of the closures of the fragments, so that a sentence is true according to the corpus if and only if it is true according to every fragment; that satisfies all the desiderata except (1). What I doubt is that there is any useful and intuitive conception that satisfies all five desiderata. When asked to respect (1)–(4), I come up with a conception that violates (5). But a natural conception that satisfies all five is what the relevantist needs, if he is to vindicate relevance in the way we have been considering.

III

I conclude that if we seek a vindication of relevance, we must find a third way in which sentences can be regarded as both true and false. I suggest we look to ambiguity. (Similar suggestions appear in Dunn [5]: 167–9 and in Pinter [9]. But I am not sure that these authors mean what I do by ambiguity. Dunn illustrates it with what seems more like a case of unspecificity. Pinter defines an ambiguous proposition as one that is both true and false; if that applies to sentences at all, it makes "Fred went to the bank" unambiguous if he went nowhere, or if he went to the riverside branch of the First National.)

Strictly speaking, an ambiguous sentence is not true and not false, still less is it both. Its various disambiguations are true or false *simpliciter*, however. So we can say that the ambiguous sentence is true or is false on one or another of its disambiguations. The closest it can come to being simply true is to be true on some disambiguation (henceforth abbreviated to "osd"); the closest it can come to being simply false is to be false-osd. Barring independently motivated truth-value gaps for the disambiguations, we have just three possibilities. A sentence can be true-osd only, false-osd only, or both true-osd and false-osd. That is, it can be true on all its disambiguations, false on all, or true on some and false on others. Here we have a possible intuitive interpretation of the three-valued semantics of Versions RM and LP.

It would be hard to get the four-valued semantics of Version E. We would be trying for the fourth value: neither true-osd nor false-osd. That makes sense if we have truth-value gaps; a sentence might be gappy on all its disambiguations. (Presumably any sentence has at least one disambiguation.) But if we have gaps, how do we end up with four values rather than seven? I can see no reason to exclude these three extra values:

gappy-osd, true-osd, not false-osd;
gappy-osd, not true-osd, false-osd;
gappy-osd, true-osd, false-osd.

Accordingly, I shall consider the three-valued versions only.

We must check that the orthodox rules for the connectives still hold when truth and falsity are understood as truth-osd and falsity-osd. They do, provided that we admit *mixed disambiguations:* that is, disambiguations of a compound sentence in which different occurrences of the same ambiguous constituent are differently disambiguated. We should admit them; they are commonplace in ordinary language. Consider the most likely disambiguation of "Scrooge walked along the bank on his way to the bank" – he fancied a riverside stroll before getting to work with the money.

It makes no sense to say that an implication involving ambiguous sentences preserves truth *simpliciter.* But it may preserve truth-osd. Also it may preserve truth-osd-only, in other words truth on all disambiguations. The implications that preserve truth-osd are those given by the first-degree fragment of Priest's LP. Those that preserve both truth-osd and truth-osd-only are given by the first-degree fragment of R-mingle. So we have two logics for ambiguous sentences – and lo, they are partly relevant.

Logic for ambiguity – who needs it? I reply: pessimists.

We teach logic students to beware of fallacies of equivocation. It would not do, for instance, to accept the premise $A \lor B$ because it is true on one disambiguation of A, accept the premise $\sim A$ because it is true on another disambiguation of A, and then draw the conclusion B. After all, B might be unambiguously false. The recommended remedy is to make sure that everything is fully disambiguated before one applies the methods of logic.

The pessimist might well complain that this remedy is a counsel of perfection, unattainable in practice. He might say: ambiguity does not stop with a few scattered pairs of unrelated homonyms. It includes all sorts of semantic indeterminacy, open texture, vagueness, and whatnot, and these pervade all of our language. Ambiguity is everywhere. There is no unambiguous language for us to use in disambiguating the ambiguous language. So never, or hardly ever, do we disambiguate anything fully. So we cannot escape fallacies of equivocation by disambiguating everything. Let us rather escape them by weakening our logic so that it tolerates ambiguity; and this we can do, it turns out, by adopting some of the strictures of the relevantists.

The pessimist may be right, given his broad construal of "ambiguity" (which is perfectly appropriate in this context), in doubting that full disambiguation is feasible. But that is more of a remedy than we really need. For purposes of truth-functional logic, at least, it is good enough if we can disambiguate to the point where everything in our reasoning is true-osd only or false-osd only, even if ambiguity in sense remains. Can we do as well as that?

Certainly we often can; maybe we always can. Or maybe we sometimes can't, especially in perplexing subject matters where ambiguity is rife. I have no firm view on the question. If indeed we sometimes cannot disambiguate well enough, as the pessimist fears, then it may serve a purpose to have a partly relevant logic capable of stopping fallacies of equivocation even when equivocation is present.

There is indeed a sense in which classical logic preserves truth even in the presence of ambiguity. If an implication is classically valid, then for every unmixed disambiguation of the entire implication, in which each ambiguous constituent is disambiguated the same way throughout all the premises and the conclusion, the conclusion is true on that disambiguation if the premises are. But although this gives a clear standard of validity for ambiguous implications, the pessimist will doubt that it is a useful standard. If things are as bad as he fears, he must perforce reason from premises accepted merely as true-osd, or at best as true-osd only. If he cannot disambiguate further, how can he tell whether his premises are made true together by any unmixed disambiguation? His highest hope for his conclusion is that it will be true-osd only, or at least true-osd. But he cannot tell whether those properties carry over from his premises to his conclusion just by knowing that unmixed disambiguations of the entire implication always preserve truth.

Jaśkowski [6] put forward his "discursive logic" *inter alia* as a logic for ambiguity. His standard of validity for ambiguous implications differs from those we have considered. He requires unmixed disambiguation of each sentence that occurs as a premise or conclusion, but allows an ambiguous constituent to be disambiguated differently in different sentences of an implication. If an implication is valid by Jaśkowski's standard, then if each premise is true on some unmixed disambiguation, the conclusion is so as well. The pessimist will complain that this standard too is useless, though better than the classical one. He knows that an implication satisfies Jaśkowski's standard; he accepts each premise as true-osd, or even as true-osd only; but he still cannot tell whether to accept the conclusion as true-osd, unless he can tell whether the disambiguations that make his premises true are mixed or unmixed. But how is he to tell that if he cannot disambiguate the premises?

REFERENCES

[1] Alan Ross Anderson and Nuel D. Belnap, Jr., "Tautological Entailments," *Philosophical Studies* 13(1962): 9–24.

[2] Nuel D. Belnap, Jr., "How a Computer Should Think," in G. Ryle, *Contemporary Aspects of Philosophy* (Oriel Press, 1977).

[3] J. Michael Dunn, "Intuitive Semantics for First-Degree Entailments and 'Coupled Trees'," *Philosophical Studies* 29(1976): 149–68.

[4] ———, "A Kripke-Style Semantics for R-Mingle Using a Binary Accessibility Relation," *Studia Logica* 35(1976): 163–72.

[5] ———, "A Theorem in 3-Valued Model Theory with Connections to Number Theory, Type Theory, and Relevant Logic," *Studia Logica* 38(1979): 149–69.

[6] Stanisław Jaśkowski, "Propositional Calculus for Contradictory Deductive Systems," *Studia Logica* 24(1969): 143–57.

[7] D. C. Makinson, *Topics in Modern Logic* (Methuen, 1973).

[8] Robert K. Meyer, "A Boolean-Valued Semantics for R," Research Paper No. 4 of the Logic Group, Department of Philosophy, Research School of Social Sciences, Australian National University.

[9] Charles Pinter, "The Logic of Inherent Ambiguity," in A. I. Arruda, N.C.A. Da Costa, and A.M. Sette, *Proceeding of the Third Brazilian Conference on Mathematical Logic* (Sociedade Brasileira de Lógica and Universidade de São Paulo, 1980).

[10] Graham Priest, "The Logic of Paradox," *Journal of Philosophical Logic* 8(1979): 219–41.

[11] Nicholas Rescher and Robert Brandom, *The Logic of Inconsistency* (Blackwell, 1980).

[12] Richard Routley, "Ultralogic as Universal?" *Relevance Logic Newsletter* 2(1977): 50–90 and 138–75; reprinted in Routley, *Exploring Meinong's Jungle and Beyond* (Australian National University, 1980).

[13] ———, "Dialectical Logic, Semantics and Metamathematics," *Erkenntnis* 14(1979): 301–31.

[14] ——— and Valerie Routley, "The Semantics of First Degree Entailment," *Noûs* 6(1972): 335–59.

[15] P. K. Schotch and R. E. Jennings, "Inference and Necessity," *Journal of Philosophical Logic* 9(1980): 327–40.

8

Relevant implication

In [4] I offered an analysis of what it means to be (entirely) about a subject matter. I first repeat that analysis. Then I define several relations of relevance, for instance, between the premise and conclusion of an implication. I show that whenever a premise implies a conclusion, in the ordinary sense of truth-preservation, then also the premise is relevant to the conclusion in the sense of the present analysis. *Pace* Anderson and Belnap [1], there can be no such thing as a truth-preserving "fallacy of relevance". Finally I remark that this does not by any means do away with all motivations for relevant logic.

SUBJECT MATTERS

We can think of a subject matter, sometimes, as a part of the world: the 17th Century is a subject matter, and also a part of this world. Or better, we can think of a subject matter as a part of the world *in intension:* a function which picks out, for any given world, the appropriate part – as it might be, that world's 17th Century. (If for some reason the world had no 17th Century, the function would be undefined.) We can say that two worlds are exactly alike with respect to a given

First published in *Theoria* 54 (1988), 161–174. Reprinted with kind permission from *Theoria*.

I thank John Burgess (Princeton), B. J. Copeland, Allen Hazen, and Graham Priest for valuable discussion. I thank Harvard University for research support under a Santayana Fellowship during part of the time this paper was written.

subject matter. For instance two worlds are alike with respect to the 17th Century iff their 17th Centuries are exact intrinsic duplicates (or if neither one has a 17th Century).

This being exactly alike is an equivalence relation. So instead of thinking of a subject matter as a part of the world in intension, we can think of it instead as the equivalence relation. This seems a little artificial. But in return it is more general, because some subject matters – for instance, demography – do not seem to correspond to parts of the world. Or if they do, it is because some contentious theory of "abstract" parts of the world is true.

The equivalence relation on worlds partitions the worlds into equivalence classes. The equivalence classes are propositions, ways things might possibly be. An equivalence class is a maximally specific way things might be with respect to the subject matter. So a third way to think of a subject matter, again general, is as the partition of equivalence classes.

We can associate the partition with a question, more or less as in [2]. The partition gives all the alternative complete answers to the question; the question asks which cell of the partition is the true one. (Which cell does our world fall into?) So a fourth way to think of a subject matter, again general, is as a question. The other way around, we can think of some questions as taking the form: what is the whole truth about such-and-such subject matter? (We often ask easier questions, of course. We demand not the whole truth, but only, say, a paragraph-length answer that hits the highlights.) Sometimes the best way to denote a subject matter is by a clause derived from a question. One subject matter is the question: how many stars there are.

INCLUSION OF SUBJECT MATTERS

A big subject matter, the 17th Century, includes the smaller, more specialized subject matter, the 1680's. A big subject matter, how many stars there are, includes the smaller subject matter, whether there are finitely or infinitely many stars.

When we think of subject matters as parts of the world, or rather parts of the world in intension, we can say that M *includes* N iff, for each world w, N(w) is part of M(w) (or both are undefined, or N(w) alone is undefined). This definition will apply to the 17th Century and the 1680's; not, or not in any obvious and uncontentious way, to how many stars there are and whether there are finitely or infinitely many. The definitions to follow will apply generally.

When we think of subject matters as equivalence relations, we can say that M includes N iff whenever M(v,w) then also N(v,w). If two worlds are alike with respect to the bigger subject matter, *a fortiori* they are alike with respect to the smaller. It is easier to be alike with respect to the smaller subject matter, so more worlds manage to do the easier thing. (Caution! – if M is more inclusive than N *qua* subject matters, then N is more inclusive than M *qua* equivalence relation, i.e. *qua* set of ordered pairs. I take it that when we talk of "inclusion" of subject matters, then we are speaking not literally but analogically; not of the genuine relation of part to whole, but of a relation that formally imitates it.)

When we think of subject matters as partitions, we can say that M includes N iff every cell of N is a union of cells of M. The bigger subject matter is a refinement of the smaller. A class of worlds all alike with respect to the smaller subject matter may yet subdivide with respect to the bigger.

ABOUTNESS

A proposition is *about* a subject matter, and it is a subject matter *of* the proposition, iff the truth value of that proposition supervenes on that subject matter. When we think of subject matters as equivalence relations, we can say that P is about M iff, whenever M(w,v), both w and v give P the same truth value. Contrapositively, if two worlds give the proposition different truth values although they are entirely alike with respect to a subject matter, then that proposition cannot have been entirely about that subject matter. When we think of subject matters as partitions, we can say that P is about M iff each cell of M either implies or contradicts P. The cells are maximally specific propositions about the subject matter, and accordingly must imply or contradict any other proposition about the same subject matter.

(If a proposition is, in some sense, partly about one subject matter and partly about another, then we would not expect its truth value to supervene on either one. Our supervenience definition, therefore, defines what it means to be *entirely* about a subject matter. For some senses of partial aboutness, see [4].)

LEAST SUBJECT MATTERS

If P is about M, then also P is about any subject matter that includes M. So we can speak of *a* subject matter of P, but not unequivocally of *the* subject matter of P. The closest we can come is to define the *least* subject matter of P as that subject matter M, if such there be, such that M is a subject matter of P, M is included in any other subject matter of P, and there is no other subject matter of which the same is true.

For any proposition P, we have a question: whether or not P. If this question is a genuine subject matter, then we can say that P is entirely about M iff M includes the subject matter: whether or not P. And if so, then also the least subject matter of P will be the question: whether or not P.

But sometimes we might well decline to count the question whether or not P as a genuine subject matter. For instance, when P is noncontingent, that question is a thoroughly *degenerate* subject matter. It is the question whether the necessary proposition or the impossible proposition is true, or in other words the universal equivalence relation on worlds, the one-celled partition. A degenerate subject matter, or not a genuine subject matter at all? This is a verbal question only. I choose the second alternative for convenience later. We may want to be restrictive in other ways too, for instance in excluding unduly gerrymandered, unnatural subject matters. Whenever we decide, for whatever reason, not to count the question whether P as a genuine subject matter, then P may or may not turn out to have something else as its least subject matter. It may turn out, therefore, that a proposition has subject matters, but has no least subject matter.

NONCONTINGENT PROPOSITIONS

When P is a noncontingent proposition, necessary or impossible, probably it has no least subject matter (given that we don't count the degenerate subject matter just mentioned). But it does have subject matters. It turns out that a noncontingent proposition is about *any* subject matter M, since whenever M(w,v) then w and v give it the same truth value. This is an immediate consequence of our definition of aboutness as supervenience.

It is a surprising consequence, no doubt. But there is more to say in its favor. First, for the case that P is necessary. Then P is contentless; it gives us no information about anything, it rules out nothing at all. Now observe that our conception of aboutness as supervenience is a double-negative notion. A proposition is entirely about a subject matter iff *none* of its content is *not* about that subject matter. We do not require complete coverage, rather we forbid straying outside. A proposition about turkeys is entirely about poultry even if it gives us no information about chickens, for it gives us no information about anything else but poultry. But a contentless proposition does better still. It gives us no information about anything, *a fortiori* no information about non-poultry. It never strays anywhere, thereby it keeps inside whatever boundary you please.

The same cannot be said when P is impossible. Then P seems to give us altogether too much information, ruling out all possibilities for all subject matters, straying out of all bounds. However, in all other cases, it is intuitively compelling that a proposition and its negation should be exactly alike with respect to what they are about; and in all cases including this, a proposition and its negation supervene on exactly the same subject matters. Therefore it is best, on the whole, to say that an impossible proposition also is about the same subject matters as its denial; and its denial is necessary, and so is vacuously about every subject matter.

We have four conflicting intuitive *desiderata*. (1) Impossible propositions seem never to be entirely about subject matter M (unless it be the greatest subject matter, the equivalence relation on which no two worlds are equivalent). (2) Necessary propositions seem always to be entirely about any subject matter M, since they never give any information not about M. (3) Necessary propositions and their impossible negations should come out alike in subject matter. Finally, (4) noncontingent propositions should not get special treatment, but should fall in line with the contingent propositions, for which aboutness as supervenience works smoothly. We can't have all four. Respecting (1) and (2) together loses both (3) and (4). Preferring (1) to (2) saves (3) but still loses (4). We do best to drop (1) and save the rest.

You may well protest that we have a deeper problem. Even granted that necessary propositions should be about the same subject matters as their impossible negations, it still seems that we ought to be able to distinguish the different subject matters of different necessary propositions, and likewise of different impossible propositions. (In which case we must first admit more than just two noncontingent propositions, one necessary and one impossible. No worries; many conceptions of propositions are available, some intensionally and some hyperintensionally individuated, and — as always — we should choose a conception that suits the job at hand.) Pity the poor mathematics librarian! How can he classify his books? Some are about number theory, some are about topology, . . . , and yet their propositional content is noncontingent through and through. Now our definition of aboutness as supervenience may seem bankrupt. Not so. Before we are done, we shall find out how to solve the librarian's problem. But for now, I accept your protest as fair, leave it unanswered, and move on.

MEREOLOGY OF SUBJECT MATTERS

Once we know what it means to say that one subject matter includes another, we can also say that two subject matters *overlap* (again, in a non-literal and analogical sense) iff they have some subject matter as a common part, included in both. Otherwise they are *disjoint*. We can define the *sum* of subject matters M_1, M_2, . . . as the subject matter, if such there be, that includes all of the M's, and is included in any other subject matter of which the same is true. We can define the *intersection* of subject matters M_1, M_2, . . . as the subject matter, if such there be, that is included in all the M's, and includes any other subject matter of which the same is true.

(If we had counted the degenerate subject matter — the universal relation on worlds, the one-membered partition — it would have been included in every subject matter. So in a trivial way, all subject matters would have overlapped. It was to avoid this, and to avoid having to talk always of "non-trivial overlap" instead of overlap *simpliciter*, that I chose not to count the degenerate subject matter.)

ORTHOGONALITY AND CONNECTION

Two subject matters M_1 and M_2 are *orthogonal* iff, roughly, any way for M_1 to be is compatible with any way for M_2 to be. If we think of subject matters as equivalence relations, orthogonality means that for any worlds w and v there is a world u such that $M_1(u,w)$ and $M_2(u,v)$. If we think of subject matters as partitions, orthogonality means that M_1 and M_2 cut across each other: each cell of M_1 intersects each cell of M_2. Subject matters are *connected* iff they are not orthogonal.

If M_1 is included in N_1, and M_2 is included in N_2, then if M_1 and M_2 are connected, so are N_1 and N_2. *Proof.* We show the contrapositive. If N_1 and N_2 are orthogonal, then for any w and v, we have u such that $N_1(u,w)$ and $N_2(u,v)$. By the two inclusions, also $M_1(u,w)$ and $M_2(u,v)$, so M_1 and M_2 also are orthogonal. *QED*

Whenever two subject matters overlap, they are connected. *Proof.* Suppose not. Then we have N included in M_1 and M_2, which are orthogonal. Let W and V be two equivalence classes of N (since N is non-degenerate). Let w be in W, and v in V; then we have u such that $M_1(u,w)$ and $M_2(u,v)$. Because N is included in M_1 and M_2, $N(u,w)$ and $N(u,v)$; so $N(w,v)$, contradicting the supposition that w and v fall in different equivalence class of N. *QED*

(What about the converse? If arbitrary non-degenerate equivalence relations may count as subject matters, then there are subject matters that are connected without overlapping. *Example:* eight worlds like this, where dotted lines indicate M_1-equivalence and dashed lines indicate M_2-equivalence.

```
0----0
.    .
.    .
.    .
0----0----0
.    .    .
.    .    .
.    .    .
0----0----0
```

But I find it hard to think of a natural example of connection without overlap. Maybe such cases ought to be excluded by a constraint on which equivalence relations count as genuine subject matters.)

RELEVANCE OF PROPOSITIONS

We can define four relations of relevance between propositions, as follows.

Identity: P and Q have the same least subject matter; more generally, all and only the subject matters of P are subject matters of Q.

Inclusion: the least subject matter of P is included in the least subject matter of Q; more generally, some subject matter of P is included in all subject matters of Q; equivalently, every subject matter of Q is a subject matter of P.

Overlap: the least subject matter of P overlaps the least subject matter of Q; more generally, every subject matter of P overlaps every subject matter of Q.

Connection: the least subject matter of P is connected with the least subject matter of Q; more generally, every subject matter of P is connected with every subject matter of Q.

Identity implies inclusion; overlap implies connection. Inclusion, or even identity, does not imply overlap; for a counterexample, suppose M and N are subject matters of both P and Q, but no common part of M and N counts as a genuine subject matter. Inclusion, or even identity, does not imply connection; for a counterexample, let one or both of P and Q be noncontingent. But for contingent P and Q, inclusion (or identity) does imply connection. *Proof.* P and Q are relevant by inclusion in one or the other direction; let it be that every subject matter of Q is a subject matter of P. Since P is contingent, P is true at a world w and false at a world v. Suppose for *reductio* that some subject matter M_1 of P is orthogonal to some subject matter M_2 of Q. M_2 is a subject matter also of P. There is a world u such

118

that $M_1(u,w)$ and $M_2(u,v)$; but then P is both true and false at w, which is impossible. So P and Q are connected. *QED*

Let us say that two propositions are *relevant* to one another iff at least one of the four relations of relevance holds between them; equivalently, iff either inclusion or connection holds between them; equivalently, iff either connection holds between them or else one of them is noncontingent.

IMPLICATION

Proposition P *implies* proposition Q iff every world that makes P true makes Q true as well.

Whenever P implies Q, P and Q are relevant. *Proof.* If either P or Q is noncontingent, it is about all subject matters, in which case P and Q are relevant by inclusion in one or the other direction. Otherwise, P is true at some world w and Q is false at some world v. If P and Q were not relevant, then some subject matter M_1 of P would have to be orthogonal to some subject matter M_2 of Q. Then there would be a world u such that $M_1(u,w)$ and $M_2(u,v)$; so P would be true and Q false at u, contradicting the implication of P by Q. *QED*

We can extend the result: whenever either P or its negation implies either Q or its negation, P and Q are relevant. *Proof.* A proposition and its negation must supervene on exactly the same subject matters, hence must stand in exactly the same relations of relevance. *QED*

What of the converse? If neither P nor its negation implies either Q or its negation – if P and Q are *logically independent* – and if the question whether or not P and the question whether or not Q are genuine subject matters, then these two subject matters are orthogonal; and further, P and Q are contingent; whence it follows that P and Q are not relevant. But if we reject these questions as genuine subject matters, then it may happen that P and Q are relevant despite their independence.

Example. Let P be the proposition that Fred is either in Carneys Point or in Ellerslie or in Germiston or in Mundrabilla or in Northcote; and let Q be the proposition that Fred is either in Ellerslie or in Heby or in Noke or in Northcote or in Zumbrota. P and Q are logically independent. If the two orthogonal questions whether P and whether Q are genuine subject matters, then P and Q turn out not to be relevant – which may seem wrong. It may well seem better, then, not to count the two questions as genuine subject matters, due to the miscellaneously disjunctive character of P and Q. A much more natural subject matter is close at hand: Fred's whereabouts. This is a subject matter of P and of Q; perhaps it is the least among the genuine subject matters of the two propositions, once we throw away the gerrymanders. If so, then P and Q, despite their independence, are relevant by identity.

ARGUMENT-FORMS

An *argument-form* is a schema for a pair of sentences. (We disregard multi-premise arguments; let the premises be conjoined.) It is *truth-preserving* iff, whenever a pair of sentences is an instance of it, the proposition expressed by the first implies the proposition expressed by the second. It is *relevant* iff, whenever a pair of sentences is an instance of it, the proposition expressed by the first is relevant to the proposition expressed by the second. Some argument-forms are neither truth-preserving nor relevant; these are fallacies of relevance. *Example:* "If you do not agree that A, I shall beat you with this stick. Therefore A." Some argument-forms are relevant but not truth-preserving. These are not fallacies of relevance, but they are still fallacies. *Example:* "A. Therefore not-A." Since whenever one proposition implies another the two are relevant, every truth-preserving argument-form is also relevant. A Pittsburgh "fallacy of relevance," as denounced in [1], would be an argument form that was truth-preserving but not relevant; according to the present treatment, there are none of those.

Ex falso quodlibet (for short, *quodlibet*) is the argument-form "A and not-A. Therefore B." It is truth-preserving, since in every instance of it, the contradictory premise expresses the impossible proposition. Since the impossible proposition is about every subject matter, *a fortiori* it is about every subject matter of the proposition expressed by the conclusion, so we have relevance by inclusion.

Quodlibet is supposed to be the Pittsburgh "fallacy of relevance" *par excellence,* but according to the present treatment, it can be noth-

120

ing of the kind. The very last place to look for irrelevance will be an argument-form where either the premise or the conclusion is either a contradiction or a tautology.

RELEVANT LOGIC

It may seem by now that I am bashing relevant logic. Not really. Despite appearances, relevant logic is not an ideological movement, or even a philosophical position. It is a certain body of technology that can be applied in the service of quite a wide range of different philosophical positions – some more interesting than others, some more plausible than others. For one safe and dull application, see [5]; for a daring and interesting one, see [6]. (The technology may have applications outside philosophy, for instance to efficient automated reasoning.) It is not even very helpful to say that the philosophical positions served by relevant logic are united by a common animus against *quodlibet*. The complaints lodged against *quodlibet* are just too varied: it is truth-preserving but irrelevant, it is not truth-preserving in paradoxical cases, it does not preserve truth-according-to-a-story, it does not preserve truth-on-some-disambiguation, it is not a way we actually reason, it would be dangerous if we did reason that way, it just is not what we call "implication."

I mean to bash only one of all the motivations for relevant logic, only one of all the complaints against *quodlibet:* namely, the idea that a good argument-form must have two separate virtues, truth-preservation and relevance, and *quodlibet* has the first but lacks the second.

We saw how to make the case that *quodlibet* is as relevant an argument-form as any, and indeed that every truth-preserving argument-form is relevant. The trivial way that *quodlibet* is relevant is very like the trivial way that it preserves truth. The proposition expressed by a contradiction is about any subject matter because, since there is no way at all for two worlds to give it different truth values, *a fortiori* there is no way for two worlds to give it different truth values without differing with respect to the subject matter. An argument from a contradiction preserves truth because, since there is no way at all for a world to make the premise true, *a fortiori* there is no way for a world to make the premise true without making the conclusion true.

I do not say that you have to like these two arguments, or that you cannot resist them. Far from it! But, given their similarity, it seems highly arbitrary to resist the first and accept the second. The reason why *quodlibet* is relevant and the reason why it is truth-preserving go together like hand and glove.

If you want to resist both arguments, you know what to do. (See [3] pp. 40–44, [6], [7], [8].) Extend the class of worlds to a broader class of *worlds**, as we may call them: ways, not necessarily possible, for things to be. A *proposition** is a class of worlds*. It is *non-contingent** iff it has the same truth value at all worlds*; *necessary** if everywhere true, *impossible** if everywhere false. One proposition* *implies** another iff no world* makes the first true without making the second true. Extend semantics so that a sentence *expresses** a proposition*. We can think of a *subject matter** now as an equivalence relation extended from worlds to worlds*. A proposition* P is *about** a subject matter* M iff, for any two worlds* w and v, if M(w,v) then w and v give P the same truth value. Two subject matters* M_1 and M_2 are *connected** iff it is not the case that for any worlds* w and v there is a world* u such that $M_1(u,w)$ and $M_2(u,v)$. One proposition* is *relevant** to another iff either every subject matter* of the first is a subject matter* of the second, or *vice versa,* or else any subject matters* of the two are connected*; equivalently, iff either any subject matters* of the two propositions* are connected* or else at least one of the two propositions* is noncontingent*. An argument-form is *truth-preserving**, or is *relevant**, iff, in any instance of it, the proposition* expressed* by the premise implies*, or is relevant* to, the proposition* expressed* by the conclusion.

Splattering stars all over the page cannot affect our argument. Just as before, a noncontingent* proposition* is about* every subject matter*. An impossible* proposition* implies*, and also is relevant* to, every proposition*. Whenever one proposition* implies* another, then the first is relevant* to the second. Any truth-preserving* argument-form is relevant*.

None of that changes. But it is up for grabs how it applies to the arguments we take for instances of *quodlibet*. If a proposition* expressed by some sentence we take for a contradiction is not impossible* but merely impossible, false at all worlds but true at some paradoxical worlds*, then there is no reason why this proposition* should either imply* or be relevant* to every proposition*. If we count such cases as genuine instances of *quodlibet* (or better yet, if *all* instances of *quodlibet* are of this kind), then it turns out that *quodlibet* is neither relevant nor truth-preserving. Now the reason why *quodlibet* is irrelevant and the reason why it is not truth-preserving go together like hand and glove. Again, it would be arbitrary in the extreme to say that it is one but not the other.

It is also up for grabs whether the content of the books in the mathematics library is contingent*, despite not being contingent. If it is, the librarian may carry on classifying them without in any way departing from a supervenience conception of aboutness*.

(But his troubles may not yet be at an end. For someday he may have to classify some books on mathematics*: a subject which parallels ordinary mathematics, and which is noncontingent* in the same way that mathematics itself is noncontingent. When that day comes he may hope to discover a broader class of worlds**. . . .)

Needless to say, the undoing of *quodlibet* and the distinguishing of noncontingent subject matters are both of them easier said than done. What I just said is somewhere between "pure" and "applied" – a wave of the hand in the direction of many very different philosophical applications of relevant logic. To finish the job, we would be obliged to take up several questions.

(1) What is the nature of the worlds*? Are they the same sort of things as the actual world, the big thing consisting of us and all our surroundings? Or are they mathematical constructions out of parts of the actual world? Or out of parts of all the possible worlds? Do we want to allow only subtly, as opposed to blatantly, impossible worlds*? If so, how do we draw the line?

(2) How do we extend semantics so that sentences not only express propositions, but also express* propositions*? Is this extension arbitrary, or is it governed by pre-existing meanings? (If one wishes to criticize earlier views or comment on ordinary-language arguments, it had better be the latter!)

(3) How do we extend the equivalence relations of subject matters so that they partition worlds*? Again, is this extension arbitrary?

(4) Finally, why should argument-forms be evaluated in terms of truth-preservation* and relevance*?

Different appliers of relevant logic, with different philosophical views, can be expected to undertake this agenda in very different ways.

REFERENCES

[1] Anderson, A. R., and Belnap, N. D. *Entailment: the logic of relevance and necessity,* vol. 1. Princeton: Princeton University Press, 1975.

[2] Belnap, N. D., and Steel, T. B. *The logic of questions and answers.* New Haven: Yale University Press, 1976.

[3] Cresswell, M. J. *Logics and languages.* London: Methuen, 1973.

[4] Lewis, D. "Statements partly about observation." *Philosophical papers,* vol. 17 (1988), pp. 1–31; reprinted as Chapter 9 of this volume.

[5] Lewis, D. "Logic for equivocators." *Noûs,* vol. 16 (1982), pp. 431–41; reprinted as Chapter 7 of this volume.

[6] Priest, G. *In contradiction.* Dordrecht: Nijhoff, 1987.

[7] Routley, R. and V. "The semantics of first degree entailment." *Noûs,* vol. 6 (1972), pp. 335–59.

[8] Routley, R., Meyer, R. K., Plumwood, V., and Brady, R. T. *Relevant logics and their rivals,* Part I. Atascadero, California: Ridgeview, 1982.

9

Statements partly about observation

Some statements are entirely about observation. An uncompromising empiricist might say that these statements alone are meaningful; but in that case, theoretical science shares in the downfall that was meant for metaphysics. An uncompromising empiricist might tough it out: science is indeed meaningless, but yields meaningful theorems; or it is entirely about observation, after all; or some of each. But it seems, rather, that science is partly about observation and what we can observe, and partly about the hidden causes and minute parts of what we can observe. And it seems also that science is a package deal, which cannot credibly be split into one part that is meaningful and one part that isn't.

The sensible empiricist, therefore, will retreat. Statements entirely about observation may remain at the core of the meaningful, but scientific statements also will be admitted. Collectively, and even individually, these are at least partly about observation. For an empiricist who wants to be a friend to science, that had better be good enough.

I

One empiricist who sought to eliminate metaphysics but spare science was A.J. Ayer.[1] Meaningful statements need not be entirely about observation.

First published in *Philosophical Papers* 17 (1988), 1–31. Reprinted with kind permission from *Philosophical Papers*.
1 *Language, Truth and Logic* (London: Gollancz, 1936; second edition, 1946). Citations are to the second edition.

. . . the question that must be asked about any putative statement of fact is not, Would any observations make its truth or falsehood logically certain? but simply, Would any observation be relevant to the determination of its truth or falsehood? And it is only if a negative answer is given to this second question that we conclude that the statement under consideration is nonsensical. (p. 38)

Dissatisfied with this use of the notion of evidential relevance, he offers a 'clearer' formulation.

Let us call a proposition which records an actual or possible observation an experiential proposition. Then we may say that it is the mark of a genuine factual proposition . . . that some experiential propositions can be deduced from it in conjunction with certain other premises without being deducible from those other premises alone. (pp. 38–39)

The criterion collapses: any statement whatever turns out to be either 'factual' or 'analytic', and meaningful in either case. Let S be any statement and let O be an experiential proposition. Then O follows from S in conjunction with the premise 'if S then O'; and thereby S qualifies as factual unless O follows from the premise alone. But O follows from 'if S then O' just when O follows from 'not S'. So if S is not factual, every experiential proposition must follow from 'not S'; and in that case, given the safe assumption that some two experiential propositions are incompatible, S must be analytic.

In his introduction to the second edition, Ayer notes the collapse.[2] (p. 11) Therefore he emends the criterion, and it is this second try that I shall be discussing henceforth.

I propose to say that a statement is directly verifiable if it is either itself an observation-statement, or is such that in conjunction with one or more observation-statements it entails at least one observation-statement which is not deducible from these other premises alone; and I propose to say that a statement is indirectly verifiable if it satisfies the following conditions: first,

---

2 However, he gives an incorrect proof of it, overlooking that a conditional may imply its own consequent. See my 'Ayer's First Empiricist Criterion of Meaning: Why Does it Fail?' *Analysis* 48 (1988) 1–3; reprinted as Chapter 10 of this volume.

that in conjunction with certain other premises it entails one or more directly verifiable statements which are not deducible from these other premises alone; and secondly, that these other premises do not include any statement that is not either analytic, or directly verifiable, or capable of being independently established as indirectly verifiable. (p. 13)

A statement is meaningful, by the new criterion, if and only if it is directly or indirectly verifiable, or else analytic.

Church soon showed that the new criterion also collapses.[3] Subsequent emendations, proceeding by the one-patch-per-puncture method, have led to ever-increasing complexity and ever-diminishing contact with any intuitive idea of what it means for a statement to be empirical.[4] Even if some page-long descendant of Ayer's criterion did escape collapse, provably admitting more than the observation-statements and less than all the statements, we would be none the wiser. We do not want just any class of statements that is intermediate between clearly too little and clearly too much. We want the right class. And to understand what we want, we need more guidance than just that good science should be in but the life and times of the Absolute should be out. Therefore we might do well to return to Ayer's criterion, unpatched, and try to see better not only why it fails, but also why it seems as if it should have worked.

To that end, I introduced the story with a tendentious twist. I said that the aim was to admit as meaningful a class of statements 'at least partly about observation'. It is unlikely that the empiricist himself would state his aim in this way – certainly Ayer does not. For he might well regard the notion of aboutness as unclear and dispensable: resistant to analysis (at least in austerely logical terms), perhaps ambiguous in ways that escape notice, and therefore best avoided in any official statement of his position. But if in an unofficial mood he were willing to speak of aboutness at all, then I think he might

3 Alonzo Church, review of the second edition of *Language, Truth and Logic*, *Journal of Symbolic Logic* 14 (1949) 52–53.
4 For a history of the ups and downs in this project, see Section VII of Crispin Wright, 'Scientific Realism, Observation and the Verification Principle', in *Fact, Science and Morality: Essays on A.J. Ayer's Language, Truth and Logic*, ed. by Graham Macdonald and Crispin Wright (Oxford: Blackwell, 1986).

accept my statement of his aim. I have put words in his mouth, but they sound not out of place.

I suggest that the reason why Ayer's criterion seems as if it should have worked is that it conforms to correct principles about partial aboutness. The reason why it fails is that the principles are not correct together. 'Partly about' is indeed badly ambiguous. We can distinguish two conceptions of partial aboutness, quite different but equally worthy of the name. One of the principles built into Ayer's criterion is right for the first conception, wrong for the second. Another is right for the second, wrong for the first. By combining these conflicting principles, we get collapse.

There is also a third conception of partial aboutness. Neither of the conflicting principles is right for it. However, it is the one that fits Ayer's preliminary suggestion that we should ask whether any observation would be relevant to determining the truth or falsehood of a putative statement of fact. There is also a fourth conception, which is probably irrelevant to our present interests.

I do not venture to guess whether Ayer had thoughts of partial aboutness at the back of his mind; still less, whether he was misled by conflating three different conceptions of partial aboutness. That hypothesis may offer one neat explanation of his criterion, but surely not the only explanation and very likely not the best.

Be that as it may, I think an empiricist in search of intuitive guidance *ought* to take up the idea that the desired class of empirical statements consists of statements that are, in some sense, at least partly about observation.

And not only an empiricist. Delineating the empirical need not be a prologue to debunking the rest. You might have any of many reasons for wanting to delineate a class of statements as empirical, and needing therefore to distinguish different senses in which a statement might be partly about observation. You might, for instance, want to *oppose* the thesis that empirical statements alone are meaningful; which you could not do unless you had some idea of what it meant to be empirical.

The empiricist himself may not be in the best position to delineate the empirical. Since he thinks that beyond the empirical all is nonsense, he requires a sharp and fixed boundary between the empirical and the nonsensical. The rest of us can settle for something messier. We need not worry if our delineation of the empirical turns out to be ambiguous, relative, and fuzzy, because we do not ask it to serve also as our line between sense and nonsense. The empiricist (unless he allows the latter line also to turn out messy) must perforce be less tolerant. Therefore our success need not advance his project.

The collapse of Ayer's criterion, and then the sorry history of unintuitive and ineffective patches, have done a lot to discredit the very idea of delineating a class of statements as empirical. That is reason enough why, if we think some appropriate delineation (albeit a messy one) can after all be had, we should revisit the criterion in search of principles we can salvage as correct.

II

However, the criterion as it stands is too concise. It runs together steps that we shall need to see as based on separate principles. So we must start by transforming the criterion into an equivalent formulation. That may arouse suspicion: the criterion collapses, therefore it is equivalent to anything else that collapses. But we shall give it only a very gentle, unsurprising transformation. Then it will be fair enough to say that the principles of the new formulation were there already in the original.

We build up the class of verifiable statements stepwise. (Actually, thanks to the collapse, it turns out that there is nothing left to add after the first few steps.) The first three steps together give us Ayer's *directly verifiable* statements.

(0) Begin with the class of all observation-statements.
(1) Admit all nonanalytic conditionals of the form 'If 0_1 & . . . , then 0' in which the antecedent is a conjunction of one or more observation-statements and the consequent is an observation-statement.
(2) Admit all statements that entail previously admitted statements.

Steps (1) and (2) together replace Ayer's compressed condition that we are to admit any statement P such that P, in conjunction with one or more observation-statements O_1, . . . , entails an observation-statement O which is not deducible from O_1, . . . alone.[5] Ayer's condition admits P iff our conditions (1) and (2) together do.

> *Proof.* Left to right. Suppose that P, in conjunction with O_1, . . . , entails O, but O is not deducible from O_1, . . . alone. Then we admit the conditional 'If O_1 & . . . , then O' at step (1) because it has the proper form and is not analytic; and then we admit P at step (2) because it entails the conditional.
>
> Right to left. First case: we admit P at step (2) because it entails observation-statement O. Then *a fortiori* P in conjunction with any O_1 still entails O, and we choose O_1 to be any observation-statement from which O is not deducible. Second and third cases: we admit P at step (1) because it is a nonanalytic conditional of the form 'If O_1 & . . . , then O'; or else we admit P at step (2) because it entails some such conditional. Then, either way, P in conjunction with O_1, . . . entails O, but O is not deducible from O_1, . . . alone. QED

A further sequence of steps gives us Ayer's class of *indirectly verifiable* statements. We decompress as before: each pair of our steps corresponds to one use of Ayer's condition stated in terms of entailment with the aid of extra premises. Where Ayer speaks of premises 'directly verifiable, or capable of being independently established as indirectly verifiable' we speak rather of statements previously admitted. This has the desired effect of preventing circles in which each of two statements is admitted only because the other is, yet it allows each indirectly verifiable statement to assist in the admitting of other indirectly verifiable statements after it has itself been admitted.

---

5 I construe Ayer's 'entails' and 'deducible' to cover not only narrowly logical entailment, but also deduction with the aid of analytic auxiliary premises. Thus O is deducible from O_1, . . . iff the conditional 'If O_1 & . . . , then O' is analytic. If 'entails' were given a narrowly logical sense it could turn out – and independently of the main collapse – that the conditional counts as directly verifiable although it is analytic, and that would be contrary to Ayer's intention.

Against this construal, we note that when Ayer goes on to indirect verifiability, he takes the trouble to make explicit provision for analytic auxiliary premises. If entailment with the aid of such premises is already covered, why bother? However, I think a construal on which Ayer said something superfluous is more charitable than one on which he allowed analytic statements to count as verifiable.

(3) Admit all nonanalytic conditionals of the form 'If V_1 & . . . , then D' in which the antecedent is a conjunction of one or more previously admitted statements and the consequent is a directly verifiable statement.

(4) Admit all statements that entail previously admitted statements.

And so *ad infinitum*: from here on, all odd-numbered steps are exactly like (3) and all even-numbered steps are exactly like (4).

III

The even-numbered steps give us one guiding principle: a closure condition for the class of verifiable statements under the relation of converse entailment.

> ENTAILMENT PRINCIPLE. If any statement entails a verifiable statement, then it is itself verifiable.

The Entailment Principle has a corollary which is highly plausible in its own right:

> EQUIVALENCE PRINCIPLE. If two statements are equivalent in the sense that each entails the other, then both are verifiable if either is.

I shall henceforth use the Equivalence Principle tacitly, just by declining to distinguish equivalent statements; and I shall not count this as use of the more questionable Entailment Principle.[6]

The odd-numbered steps suggest quite a different guiding principle: closure of the class of verifiable statements under certain sorts of truth-functional composition. At first it seems that we have only a quite special case.

> SPECIAL COMPOSITIONAL PRINCIPLE. If V_1, . . . are verifiable and D is directly verifiable, then unless it is analytic, the conditional 'If V_1 & . . . , then D' also is verifiable.

6 Maybe the Equivalence Principle is already built into Ayer's notion of a statement. That depends on how broad a notion of translation he has in mind when he says

But in fact we have something a good deal more general.

SPECIAL COMPOSITIONAL PRINCIPLE, REFORMULATED.
If $V_1, \ldots$ are verifiable and D is directly verifiable, and if $T(V_1, \ldots)$ is any truth-functional compound of the V's, then unless it is analytic, the disjunction '$T(V_1, \ldots)$ or D' also is verifiable.

The two formulations are equivalent. The old formulation follows instantly from the new one. The converse takes some proving.

Proof. Fix D. Consider the condition: being such that its disjunction with D is either analytic or verifiable.

First, if P is verifiable, then P satisfies the condition. For by the old formulation, 'If P, then D' is either analytic or verifiable; if 'If P, then D' is analytic, then 'P or D' is equivalent to D, which is verifiable; and if 'If P, then D' is verifiable, then by the old formulation 'If (if P then D) then D' is either analytic or verifiable, and 'If (if P then D) then D' is equivalent to 'P or D'.

Second, if P satisfies the condition, so does its negation. For if 'P or D' is analytic, then 'Not-P or D' is equivalent to D, which is verifiable; if 'P or D' is verifiable, then by the old formulation, 'If (P or D) then D' is either analytic or verifiable, and 'If (P or D) then D' is equivalent to 'Not-P or D'.

Third, if P and Q both satisfy the condition, so does their disjunction. For by the previous case, 'Not-P' and 'Not-Q' also satisfy the condition. If 'Not-P or D' and 'Not-Q or D' both are analytic, then '(P or Q) or D' is equivalent to D, which is verifiable. If 'Not-P or D' and 'Not-Q or D' both are verifiable, then by the old formulation the conditional 'If (not-P or D) & (not-Q or D), then D' is either analytic or verifiable, and this conditional is equivalent to '(P or Q) or D'. If 'Not-P or D' is analytic and 'Not-Q or D' is verifiable, then by the old formulation the conditional 'If (not-Q or D) then D' is either analytic or verifiable, and this conditional is equivalent to '(P or Q) or D'. Likewise *mutatis mutandis* if 'Not-P or D' is verifiable and 'Not-Q or D' is analytic.

that 'any two sentences which are mutually translatable will be said to express the same statement.' (p. 8) Is equivalence an adequate standard of translation, or does Ayer mean to require something stronger?

All truth functions are generated from negation and disjunction. Therefore we conclude that any truth-functional compound of verifiable statements satisfies the condition. This goes for any directly verifiable D. QED

I claim that the Entailment and Compositional Principles are separately acceptable, but should not be mixed. If we shun all mixing (except for our tacit appeals to Equivalence) we can go no further.

But now I bend my rules: one small bit of mixing turns out to do no harm, and enables us to simplify the Compositional Principle. Assume that there is at least one observation-statement 0, and consider the contradiction '0 & not-0'. '0 & not-0' entails 0, and therefore is admitted as verifiable by the Entailment Principle; in fact, it is admitted already at step (2), and therefore is *directly* verifiable.[7] Now let $V_1, \ldots$ be verifiable and let $T(V_1, \ldots)$ be any truth-functional compound of the V's; applying the Special Compositional Principle as reformulated, '$T(V_1, \ldots)$ or (0 & not-0)' is verifiable unless it is analytic; however we can simplify by dropping the contradictory disjunct. So we get a principle that applies to all forms of truth-functional composition, and that no longer uses the distinction between direct and indirect verifiability.

COMPOSITIONAL PRINCIPLE *SIMPLICITER*. If $V_1, \ldots$ are verifiable, and if $T(V_1, \ldots)$ is any truth-functional compound of the V's, then unless it is analytic, $T(V_1, \ldots)$ also is verifiable.

To make Ayer's long story short, his verifiable statements turn out to be the class we get if we start with the observation-statements (of which we assume there is at least one) and we close both under converse entailment and under truth-functional composition.[8]

7 Is it bad to count contradictions as 'verifiable'? No: whatever the target distinction may be that we are trying to capture, we would not expect it to apply to them in any intuitive way. Let their status be settled by stipulation, guided by convenience. Our settlement is the same one that follows immediately from Ayer's formulation. And if you doubt that '0 & not-0' does entail 0, bear in mind that we are not using the maligned rule *ex falso quodlibet*; we just drop the second conjunct.

8 That is, under truth-functional composition such as to yield a statement that is not analytic. Let this qualification be understood without saying henceforth.

IV

No collapse comes from the Compositional Principle by itself. (That is why no harm was done when I mixed the principles to a limited extent in advancing from the Special Compositional Principle to the Compositional Principle *Simpliciter*.) If we start with the observation-statements and close under truth-functional composition, we do not get the class of all (non-analytic) statements.[9]

Take a miniature language as follows: we have two observation-statements, 'It's dark' and 'It's light'; they are exclusive, since 'It's not both dark and light' is analytic; but they are not exhaustive, since 'It's dark or light' is not analytic. (Twilight is acknowledged as a third possibility, but doesn't have an observation-statement of its own.) Also we have two other statements, 'The Absolute is cruel' and 'The Absolute is crafty' which are independent of the two observation-statements and of each other. We admit five new statements by applying the Compositional Principle to the observation-statements: 'It's dark or light', 'It's neither dark nor light', 'It isn't dark', 'It isn't light', and the contradictory 'It's dark and light'. But we don't admit 'The Absolute is cruel', or even such conjunctions as 'It's dark and the Absolute is cruel'.

No collapse comes from the Entailment Principle by itself; or even from the Entailment Principle applied after we have first applied the Compositional Principle. Mixing is not always fatal. In the first case, what we admit are exactly the entailers of observation-statements.[10] Likewise in the second case we admit exactly the entailers of truth-functional compounds of observation statements. So if we start with the observation-statements and close under converse entailment, we admit 'It's dark and the Absolute is cruel' because it

---

9 It may be, for all I know, that the 'observation-statements' already are closed under truth-functional composition. If they are, of course this step will add nothing new. That will be so, for instance, if the observation-statements are the same thing as the 'statements entirely about observation' to be discussed shortly; whereas it will not be so if they are the statements that can be tested fairly quickly and decisively by observation.

10 Observation-statements themselves need no separate mention: they entail themselves. Entailers of entailers of observation-statements, or entailers of entailers of

entails 'It's dark'. If we start with the truth-functional compounds of observation-statements and close under converse entailment, we also admit 'It isn't dark and the Absolute is cruel' because it entails 'It isn't dark'; we admit 'It's dark or light and the Absolute is crafty' because it entails 'It's dark or light'; and so on. But in neither case do we get the class of all statements. For instance we do not admit 'The Absolute is cruel', and we do not admit 'Either it's dark and the Absolute is cruel or it's light and the Absolute is crafty'.

But the next step is the fatal one. Suppose we begin with the observation-statements, then apply the Compositional Principle, then the Entailment Principle, then the Compositional Principle once more. This is the mixing that yields collapse. First we have 'It's dark'; then 'It isn't dark'; then we have both 'The Absolute is cruel and it's dark' and 'The Absolute is cruel and it isn't dark' (so far, so good); then the disjunction of these, which is equivalent to 'The Absolute is cruel'. And in place of 'The Absolute is cruel' we may likewise admit whatever (non-analytic) statement we like.

To state the point in general form, forsaking our miniature example, let us suppose as Ayer implicitly does that the class of verifiable statements is closed both under converse entailment and under truth-functional composition. Assume that we have at least one verifiable statement V and, further, that V is not contradictory. (In other words, 'Not-V' is not analytic.) We could safely assume, for instance, that there exists at least one non-contradictory observation-statement.[11] Then any statement S whatever, unless it is analytic, is verifiable.

Proof. 'Not-V' is verifiable by the Compositional Principle. Each of 'S & V' and 'S & not-V' entails a verifiable statement and so is itself verifiable by the Entailment Principle. Then the disjunction '(S & V) or (S & not-V)', unless it is analytic, is verifiable by the Compositional

entailers of observation-statements, or . . . , need no separate mention: entailment is transitive.

11 Could we assume even less and still prove the collapse? No. The empty class is closed under converse entailment and under truth-functional composition; so without just assuming the contrary, we cannot rule out the hypothesis that no statements are verifiable. The class of contradictory statements also is closed under

Principle. The same goes for S itself, by the Equivalence Principle, since S is equivalent to this disjunction. *QED*

So much for our two principles taken together.[12]

V

Before we return to take them separately, we must explore alternative senses in which a statement might be partly about a subject matter. And before that, we must ask what it means for a statement to be entirely about a subject matter.

I suggest that this is a matter of supervenience: a statement is entirely about some subject matter iff its truth value supervenes on that subject matter. Two possible worlds which are exactly alike so far as that subject matter is concerned must both make the statement true, or else both make it false. Contrapositively, if one world makes the statement true and the other makes it false, that must be because they differ with respect to the subject matter. If the statement is entirely about the subject matter, no difference that falls outside that subject matter could make a difference to the truth of the statement.

It is simplest if we take possible worlds to be things of a kind with

converse entailment and under truth-functional composition (since we exclude composition that yields analytic statements); so without just assuming the contrary, we cannot rule out the hypothesis that exactly the contradictions are verifiable.

12 An interesting new idea for patching Ayer's criterion to avert collapse has been advanced by Crispin Wright (*op. cit.*, pp. 267–268). It invokes what I shall call *idiosyncratic entailment*.

Think of Ayer's 'statements' as sentences, so that it makes sense to speak of their syntactic constituent structure. Call X a *constituent* of a (one-premise) entailment iff X is a non-logical expression that occurs at least once in the premise. Say that substitution of Y for X *preserves* the entailment iff the result of uniformly substituting Y for X in the premise still entails the conclusion. Say that the entailment is *idiosyncratic* to X iff some substitution for X fails to preserve the entailment – the entailment works in virtue of some idiosyncrasy of X, and accordingly fails when we find a substituent for X that lacks the idiosyncrasy. Say that the entailment is *idiosyncratic* iff it is idiosyncratic to each of its constituents. (It is the opposite of a narrowly logical entailment, which is idiosyncratic to none of its constituents.)

the cosmos that we ourselves are part of,[13] and if we take a subject matter that picks out parts of some of these worlds. For instance the 17th Century is a subject matter; the thisworldly 17th Century is a

When we prove the collapse of Ayer's criterion, there is nothing idiosyncratic about it: it makes no difference whether we are admitting 'The Absolute is cruel' or 'The nothing noths' or what have you. So maybe we could stop the collapse by limiting the Entailment Principle to apply only to idiosyncratic entailment. This would be far simpler than most of the proposals to patch Ayer's criterion. Further, it would make intuitive sense – though maybe it would rest more on an intuitive conception of what counts as logical jiggery-pokery than on an intuitive conception of what counts as empirical. Try this:

The verifiable statements are the class we get if we start with the observation-statements and we close both under converse idiosyncratic entailment and under truth-functional composition.

The proposal avoids collapse, sure enough, in a sufficiently impoverished language. For instance, take a sentential language in which all the atomic sentences are independent: then there are no idiosyncratic entailments, so the verifiable statements are exactly the truth-functional compounds of observation-statements. But in a sufficiently rich language, the limitation to idiosyncratic entailment accomplishes nothing. Take a language in which, for any P and Q, we have an atomic sentence S that does not occur in P or Q, and that is equivalent to 'P & Q'. Whatever language we start with, it has a definitional extension that provides such an S for every P-Q pair. (Not to worry that it takes infinitely many definitions – we could specify them all by a single schema.) Then if P entails Q, whether idiosyncratically or not, it follows that P idiosyncratically entails S and S idiosyncratically entails Q. (Idiosyncratic entailment is not transitive.) Closure under converse idiosyncratic entailment has the same effect as closure under converse entailment *simpliciter*, except that sometimes we need two steps instead of one. Collapse ensues. Even if 'The Absolute is cruel' is not admitted as verifiable in the language we speak today, it will be admitted in the definitionally extended language we could, if we liked, speak tomorrow. Collapse brought on by definitional extension is no better than collapse straightway. The proposal fails.

(Might we say that the proposal applies only after we have replaced all defined terms by their *definientia* in primitive notation? – But a language does not come with its terms already labelled as 'primitive' or 'defined'. Then might we say that the proposal applies only after we have replaced all defin*able* terms by their *definientia*? – But if there are circles of interdefinability, that never can be accomplished.)

Wright's way of invoking idiosyncratic entailment is more complicated than the proposal just considered. But to the extent that the complications make a difference, their effect is to admit more, not less. They do nothing, therefore, to avert collapse brought on by definitional extension.

13 See my *On the Plurality of Worlds* (Oxford: Blackwell, 1986).

temporal part of this world, and likewise various otherworldly 17th Centuries are parts of various other worlds. Then two possible worlds are exactly alike with respect to the 17th Century if the 17th Century that is part of one is an exact intrinsic duplicate of the 17th Century that is part of the other (or if, for one reason or another, neither world has a 17th Century); and otherwise the two worlds differ with respect to the 17th Century. So a statement is entirely about the 17th Century iff, whenever two worlds have duplicate 17th Centuries (or both lack 17th Centuries), then both worlds give the statement the same truth value. Similarly for more scattered parts, such as the totality of all the world's styrofoam. A statement is entirely about styrofoam iff, whenever all the scattered styrofoam of one world is a duplicate of all the scattered styrofoam of the other world (or neither world contains any styrofoam), then both worlds give the statement the same truth value.[14]

It is otherwise for other subject matters. For instance, consider the subject matter: how many stars there are. Two possible worlds are exactly alike with respect to this subject matter iff they have equally many stars. A statement is entirely about how many stars there are iff, whenever two worlds have equally many stars, the statement has the same truth value at both. Maybe an ingenious ontologist could devise a theory saying that each world has its *nos-part,* as we may call it, such that the nos-parts of two worlds are exact duplicates iff those two worlds have equally many stars. Maybe – and maybe not. We

14 We get a circle here: two things are exact intrinsic duplicates iff they have exactly the same intrinsic properties; a property is intrinsic iff a statement that predicates that property of something (without introducing any extra descriptive content by its way of referring to that thing) is entirely about that thing. If you begin by accepting none of the notions on the circle, you should end still accepting none; the journey around the circle does not help you. But if you begin by accepting any, you should end by accepting all. And if you begin by half-accepting several, which I suppose to be the most likely case, then again I think you should end by accepting all. Here is one point among others where the present approach to delineating the empirical appeals to distinctions that an austere empiricist might well disdain.

I discuss the circle further in 'Extrinsic Properties,' *Philosophical Studies* 44 (1983) 197–200; and 'New Work for a Theory of Universals,' *Australasian Journal of Philosophy* 61 (1983) 343–377.

shouldn't rely on it. Rather, we should say that being exactly alike with respect to a subject matter may or may not be a matter of duplication between the parts of worlds which that subject matter picks out.

Further, even for the easy cases of the 17th Century and styrofoam, maybe some reader will take issue with my supposition that possible worlds are things of a kind with the cosmos we are part of; or with my supposition that things have scattered and disunified parts.

So it may be best, once the easy cases have shown what kind of notion of aboutness I am driving at, if we reintroduce it in a more abstract and metaphysically neutral fashion, as follows. Whatever the nature of possible worlds may be, at any rate there are many of them. With any subject matter, we can somehow associate an equivalence relation on worlds: the relation of being exactly alike with respect to that subject matter. Now, unburdened of any contentious account of what that relation and its *relata* are, we proceed as before. A statement is entirely about a subject matter, iff, whenever two worlds are exactly alike with respect to that subject matter, then also they agree on the truth value of the statement.[15]

This treatment does not, in general, give us an entity which we may naturally take to *be* the subject matter. Sometimes we have a suitable entity: we could take the subject matter styrofoam to be the totality of all the styrofoam throughout all the worlds. Then it picks out, by intersection, the styrofoam (if any) of any given world. But we cannot rely on doing the same in all cases, as witness the subject matter: how many stars there are. What we do have, in all cases, is the equivalence relation. We might dispense with subject matters as entities, and get the effect of quantifying over subject matters by quantifying instead over equivalence relations. (Perhaps over all equivalence relations on worlds; perhaps only over those which can

15 Suppose some unobvious, philosophically interesting supervenience thesis is true: perhaps the thesis that the laws of nature supervene on the spatiotemporal arrangement of local qualities. It follows that any statement entirely about the laws of nature is also entirely about the arrangement of qualities. If someone who rejects the supervenience thesis thinks he is speaking entirely about the laws of nature, and not about the arrangement of qualities, he is mistaken. This will not appeal to those who want to distance supervenience from reductionism. For myself, I welcome it. (Here I am indebted to Peter Railton.)

suitably be regarded as relations of being alike with respect to a subject matter.) Or, if we don't mind artificiality, we could simply identify a subject matter with its equivalence relation. I shall do so henceforth.

If a statement is entirely about the 1680's, then *a fortiori* it is entirely about the 17th Century; if entirely about blue styrofoam, then entirely about styrofoam; if entirely about whether there are finitely or infinitely many stars, then entirely about how many stars there are. The reason, in each case, is that the first subject matter is in some sense part of the second. In special cases, we could explain this in an especially simple way: the totality (through all the worlds) of blue styrofoam is part of the totality of styrofoam. But for the sake of generality, and to avoid contentious ontic commitments, it is better to explain part-whole relations of subject matters in terms of the equivalence relations, as follows. If two worlds are alike with respect to the entire 17th Century, then *a fortiori* they must be alike with respect to the 1680's; if alike with respect to styrofoam generally, then alike with respect to blue styrofoam; if alike with respect to how many stars there are, then alike with respect to whether there are finitely or infinitely many. In general, if subject matter M is part of a more inclusive subject matter M^+, then whenever two worlds are exactly alike with respect to M^+ – for short, M^+-*equivalent* – then they must also be M-equivalent. Identifying the subject matters with the equivalence relations: M is *part of* M^+ iff M^+ is a subrelation of M.[16] We could also say that M *supervenes on* M^+. Supervenience is transitive: when the truth value of a statement supervenes on M, and M supervenes on M^+, then the truth value of the statement supervenes on M^+. So a statement entirely about some part of subject matter M is also, *a fortiori,* entirely about M; and any statement entirely about M is also entirely about every subject matter that has M as a part.

For any subject matter M, the class of statements entirely about M is closed under truth-functional composition. If any two M-

16 This points up the artificiality of the identification. A relation is a set of pairs, a subset of a given set is part of that set; yet when M^+ is part of M in the sense of subset and set, we say that M is part of M^+ in the sense of less and more inclusive subject matter.

equivalent worlds give the same truth value to P then also they give the same truth value to 'Not-P'; if they give the same truth values to both P and Q, then also they give the same truth value to 'P & Q'; and so on.

Any two worlds whatever, and *a fortiori* any that are M-equivalent for some subject matter M, must give the same truth value to an analytic statement or a contradictory statement. In this trivial way, any analytic or contradictory statement turns out to be entirely about every subject matter. Not to worry: we should not expect distinctions of subject matter to apply in any very intuitive way to analytic and contradictory statements, so we may be content with whatever stipulation falls out of definitions that work in the cases that matter.

VI

Now take the subject matter: observation. Two worlds may or may not be exactly alike with respect to observation – for short, *observationally equivalent*. A statement is entirely about observation iff both of any two observationally equivalent worlds give it the same truth value.

It is unclear whether any part of this world, or another, may be called the totality of all the world's observation. Such a totality might be a totality of many events of observing. Some theories treat events as parts of worlds in which they occur; others do not.[17] Observational equivalence might be like the relation of having duplicate 17th Centuries, or duplicate totalities of styrofoam; or it might be more like having equally many stars. No matter; so long as it is an equivalence relation on worlds, we can go on.

You have surely spotted the vexed questions I am ignoring. Suppose two worlds look just alike to all observers, but differ because very different things are being observed. Observationally equivalent? Or suppose that in two worlds, observers respond differently not because of any difference in what stimulation they get from their surroundings, but entirely because they are primed with different preconceptions: different theory-laden concepts, different questions in

---

17 The theory in 'Events' in my *Philosophical Papers*, Volume II (Oxford: Oxford University Press, 1986) is one that does not.

mind, different training in how to observe, or just different degrees of attentiveness. Observationally equivalent? Or suppose there are two worlds where human observers are aided by instruments – maybe mere spectacles, maybe telescopes, maybe remote controlled spacecraft – and there is no difference in what ultimately reaches the humans, but plenty of difference in what reaches the instruments. Observationally equivalent? Or suppose two worlds are alike so far as the actual observations in each world go, but differ in their counterfactuals about observation. Observationally equivalent? Or.... Whenever we have questionable cases of observational equivalence, we can have questions about whether a statement is entirely about observation; because the statement might differ in truth value between worlds that are questionably equivalent, but never between worlds that are *un*questionably equivalent.

It is not my business to answer these questions. I agree, nay I insist, that the notion of observational equivalence is rife with ambiguities. Therefore, so is the notion of a statement entirely about observation. I said that we need not worry if our delineation of the empirical turns out to be ambiguous, relative, and fuzzy. It turns out that we meet ambiguity already at this stage, even before we advance from entire to partial aboutness. All this ambiguity will stay with us when we go on. But I shall disregard it henceforth. What I want to examine is the *added* ambiguity in the notion of a statement partly about observation: the ambiguity that accrues because we have several ways to go from entire to partial aboutness.

Recall that Ayer defines an observation-statement (originally, 'experiential proposition') as a statement which 'records an actual or possible observation'. It is safe to say that such a statement is a statement entirely about observation. But probably not all statements entirely about observation are observation-statements. Recall that in our miniature language we provided only two observation-statements, 'It's dark' and 'It's light' (exclusive but not exhaustive), but also we had six truth-functional compounds of these two (one analytic, one contradictory, and four more). Those six statements also are entirely about observation. Since statements entirely about observation are closed under truth-functional composition, they would seem to include statements which record not observations but *non-*

observations; not observations but very prolonged sequences of observations; not observations but conditional or biconditional correlations of observations; and so on. If such a statement were said to record *an* observation, that would be a stretch of usage, though I think not an altogether absurd stretch. At any rate, we will have statements that cannot be quickly and decisively tested by observation, and yet are entirely about observation. 'Whenever it's dark, it will later be light' is entirely about observation (if we take it to refer to observed dark and light). Yet no sequence of dawns is long enough to settle that endless night will not come at last, and no night is long enough to settle that dawn will never follow. We can restate the example with infinite conjunctions and disjunctions in place of the quantifiers, and we can approximate it with long finite ones.

VII

Now that we know, near enough, what it means to be entirely about observation, what could it mean to be (at least) partly about observation? How are we to tackle this question? Not by consulting our linguistic intuition about the ordinary use of the phrase 'partly about', I think. Because, after all, that phrase doesn't get a lot of ordinary use. Rather, we should see how the modifier 'partly' operates, and operate accordingly on the notion of being entirely about a subject matter.

The recipe for modifying X by 'partly' is something like this. Think of the situation to which X, unmodified, applies.[18] Look for an aspect of that situation that has parts, and therefore can be made partial. Make it partial – and there you have a situation to which 'partly X' could apply. If you find several aspects that could be made partial, then you have ambiguity. Maybe considerations about what it could be sensible to mean will help diminish the ambiguity.

Example. On a cloudy day, clouds cover the sky. Then what could a partly cloudy day be? Well, what in the situation has parts? First, the clouds have parts. Maybe a partly cloudy day is one on which cloud-parts cover the sky? But cloud-parts, or anyway the most salient ones, are just clouds; so there's no difference between cloud-

18 At this point you would do best to forget any technical sense of the word 'situation'.

143

parts covering the sky and clouds covering the sky; so this would be a pointless thing to mean; so it's understandable that the phrase never does mean this. Second, the day has parts. Maybe a partly cloudy day is one on which clouds cover the sky for part of the day? – Yes, the phrase can mean that. But it's still a bit pointless, since so often we could just say 'a cloudy morning' or whatever. Third, the sky has parts. Maybe a partly cloudy day is one on which clouds cover part of the sky? – Yes, and in fact this is what the phrase most often means.

When a statement is entirely about a subject matter, we have, first, the content of the statement, given by the class of possible worlds that the statement excludes. We have, second, the subject matter, given by an equivalence relation on worlds. We have, third, the supervenience of the truth value of the statement (determined by the content) upon the subject matter. And we have, fourth, the statement itself. Each of these can be taken, in some direct or some devious sense, to have parts. Therefore we have four ways to cut back from entire to partial aboutness, yielding four different conceptions of partial aboutness. I think that each of the four does indeed yield a possible meaning for the phrase 'partly about'. But whether that is so scarcely matters. What does matter is that we get four different lines of retreat from the idea that an empirical statement is entirely about observation, and three of the four can be linked to Ayer's discussion.

VIII

First, we have the *part-of-content* conception: a statement is partly about a subject matter iff part of its content is entirely about that subject matter. So far, we have been talking of aboutness for statements, not contents, but that should not detain us: if content is given by a class E of excluded worlds, E is entirely about subject matter M iff both or neither of any two M-equivalent worlds belong to E. A part of the content is a subset of E: it does part of the excluding that the whole of E does. So a statement S is partly about subject matter M, in the present sense, iff there is some subset of its content that contains both or neither of any two M-equivalent worlds.

Assume that for any content whatever, some statement has exactly that content. That could be because we have a liberal enough notion of statements to permit statements not expressible in any available language; or it could be because we have available some very rich language. Then we have simpler equivalents of the previous definition. S is partly about M, in the present sense, iff S is equivalent to a conjunction 'P & Q' where P is entirely about M and Q may be about anything. When we expand S into any equivalent conjunction, the content of each conjunct is part of the content of S; so another way to think of a part of the content of S is just to think of a conjunct of some conjunctive expansion of S. Simpler still: S is partly about M iff S entails some statement entirely about M.

For instance, in our miniature example, 'The Absolute is crafty and it's dark' is partly about observation. The part of its content that excludes it's being light or twilight is entirely about observation. The statement is equivalent (or identical) to the conjunction of 'The Absolute is crafty' and 'It's dark'; thereby it entails 'It's dark'; and 'It's dark' is entirely about observation.

S entails 'Not-0' iff 0 contradicts S; 'Not-0' is entirely about observation iff 0 is; so S is partly about observation iff some statement entirely about observation contradicts S. What we have is a liberal formulation of Falsificationism, the thesis that a statement is empirical iff it could be falsified by observation. The liberality consists in reading 'falsified by observation' as 'contradicted by a statement entirely about observation' rather than 'contradicted by an observation-statement'. That means that the falsification is not required to be at all quick and decisive.

Being partly about observation, in the sense of the part-of-content conception, obeys the Entailment Principle. (And consequently obeys the Equivalence Principle as well.) For if S_1 entails S_2, and S_2 is partly about observation, then S_2 entails some statement 0 that is entirely about observation. By transitivity S_1 also entails 0, and therefore is partly about observation.

But in return, the Compositional Principle is violated. If 0 is entirely about observation, so is 'Not-0'; then both 'S & 0' and 'S & not-0' are partly about observation. But their disjunction is equivalent to S, which might be anything. S need not be analytic, and need not be partly about observation. S might be 'The Absolute is cruel.'

145

It is false that the disjunction of two statements partly about observation, unless it is analytic, must be partly about observation.

As a delineation of the empirical, being partly about observation in the part-of-content sense seems acceptable, though I think not uniquely acceptable. As a standard of meaningfulness it is absurd; because even when part of the content is entirely about observation, the rest of the content may be about anything whatever.

IX

Second, we have the *part-of-subject-matter* conception: a statement is partly about a subject matter M iff it is entirely about a certain suitable larger subject matter M^+ which includes M as a part.

The restriction to a 'suitable' larger subject matter is essential. Without it, we could use gerrymanders to show that anything is partly about anything. We have a statement entirely about wallabies; it is therefore entirely about the larger subject matter, wallabies and tax reform; so it is partly about tax reform! As ordinary usage, that is absurd. And a conception of partial aboutness that allows it, whether ordinary or not, is so undiscriminating as to be useless.

(If we had a large mixed corpus of statements, some entirely about wallabies and some entirely about tax reform, it would not be bad to say collectively of them that they are partly about tax reform. This might be the part-of-content conception, applied to the content of the corpus as a whole. Or we might just be saying that some of the statements in the corpus are entirely about tax reform.)

The remedy is to say that the gerrymandered subject matter, wallabies and tax reform, either is no genuine subject matter at all, or else is an unsuitable subject matter for use in establishing partial aboutness. The second alternative is better, because after all we might want to say that some peculiar book is entirely about wallabies and tax reform. So we'll count it as a subject matter; but the trouble with it is that there are no salient relations between wallabies and tax reform. Everything is related to everything, of course, in countless gruesome ways. But if a subject matter is held together only by relations that we normally ignore, then that subject matter itself is best

146

ignored – at any rate for present purposes. What we want is a *close-knit* subject matter: a package deal, with its parts well interrelated in many important ways. The more close-knit the subject matter X-*cum*-Y is, the more natural it is to say that a statement entirely about X is thereby partly about Y. It seems not bad to say that a statement entirely about Buda is partly about Pest, if the life of Budapest pays no heed to the division.

It would not seem so good, however, if we also said that a statement entirely about Buda was thereby partly about each little street in Pest. So it seems we need another constraint on what is to count, for present purposes, as a 'suitable' subject matter.[19] This time, it will have to be a relative constraint: Budapest is a suitable subject matter relative to Pest, but not relative to each street in Pest. That suggests that if a statement is partly about M by being entirely about M^+, M must be a sufficiently *large* part, or a sufficiently *important* part, of M^+. Pest is a large and important part of Budapest; not so for each street in Pest. In this easy case, we can at least begin with an ordinary comparison of the sizes of material objects. In harder cases, where a subject matter does not pick out parts of worlds, we cannot. We shall have to require, in general, that the relation of M^+-equivalence does not partition the worlds too much more finely (or, too much more finely in important respects) than the relation of M-equivalence does. It would be good to spell the constraint out more exactly, but I leave that problem open.[20]

Consider the whole subject matter of science: observation, the things observed and other things of the same kind, their hidden causes and their minute parts. Call this subject matter 'observation$^+$'. Here is a larger subject matter including observation. It is eminently 'suitable'. It is well interrelated by causal relations, relations of same-

19 Here I am indebted to William Tolhurst and Terence Horgan.
20 A statement entirely about New Hampshire is entirely about New England and thereby, if New England is sufficiently close-knit and Maine is a large and important enough part of it, is partly about Maine. Nelson Goodman, in 'About', *Mind* 70 (1961) 1–24, raises this dilemma: 'Apparently we speak about Maine whenever we speak about anything contained in Maine, and whenever we speak about anything that contains Maine. But to accept this principle is to be saddled

ness of kind, and even relations of part and whole. Observation seems (so far as we can tell without spelling out a criterion exactly) to be a large and important part of it. It is a sufficiently suitable larger subject matter, I submit, that any statement entirely about it thereby qualifies as partly about observation. Of course we can say of scientific statements collectively that they are partly about observation. But an individual scientific statement also is partly about observation, even one that is entirely about 'unobservables'. Science is a package deal, observation is central to the package, and that is good enough.

Being partly about observation, in the sense of the part-of-subject-matter conception, obeys the Compositional Principle when we hold fixed the larger subject matter observation$^+$. Recall that being entirely about a given subject matter is closed under truth-functional composition. So any truth-functional compound of statements that are partly about observation by being entirely about observation$^+$ is itself entirely about observation$^+$, and thereby partly about observation. The Compositional Principle makes an exception for analytic truth-functional compounds; the exception turns out to be unnecessary, since they too will be entirely about observation$^+$ and thereby partly about observation.

The Equivalence Principle also is obeyed. If two statements are equivalent, they must supervene on exactly the same subject matters. So both or neither of them will be entirely about observation$^+$; so both or neither of them will thereby be partly about observation.

But the Entailment Principle is violated. If a statement is partly about observation by being entirely about observation$^+$, it does not follow that an entailer of that statement also is entirely about observation$^+$. 'It's dark' is entirely about observation, and *a fortiori* entirely about observation$^+$; 'It's dark and the Absolute is cruel' entails 'It's dark'; but 'It's dark and the Absolute is cruel' needn't be entirely about observation$^+$, and indeed its truth value needn't supervene on

with the conclusion that anything is about Maine.' (p. 2) He concludes that our ordinary notions concerning aboutness 'are readily shown to be inconsistent.' (p. 1) I conclude that he should have distinguished entire from partial aboutness, and the present conception of the latter from others, and suitable from unsuitable subject matters.

any suitably close-knit subject matter that includes observation as a large and important part.

As a delineation of the empirical, being partly about observation in the part-of-subject-matter sense seems acceptable when, and of course only when, we fix on a suitable larger subject matter. Whether a subject matter is suitable is, of course, a matter of degree, and a matter of judgement. The subject matter of science – observation, the things observed and other things of the same kind, their hidden causes and their minute parts – is one eminently suitable subject matter, but not necessarily the only one. Maybe some still larger subject matter might be just as suitable. If delineating the empirical means finding out what else might fall in with observation in some suitable subject matter, the task will be no mere formal exercise. Horrors! – Even the life and times of the Absolute might turn out to be partly about observation. And we could not decide without knowing just what the Absolute is supposed to be and do. This conception, like the first, cannot yield a standard of meaningfulness. We could not hope to dismiss metaphysics as meaningless before attending to its meaning.

X

Third, we have the *partial supervenience* conception: a statement is partly about a subject matter iff its truth value partially supervenes, in a suitably non-trivial way, on that subject matter. Let us say that the truth value of a statement *supervenes* on subject matter M *within* class X of worlds iff, whenever two worlds in X are M-equivalent, they give the statement the same truth value. Supervenience within the class of all worlds is supervenience *simpliciter*. Supervenience within a smaller class of worlds is partial supervenience.

The restriction to partial supervenience 'in a suitably non-trivial way' is essential. Without it, we could select classes of worlds within which anything supervenes on anything. For instance, any S supervenes on any M within the unit class of any single world; or within a class of worlds none of which are M-equivalent; or within the class of all S-worlds; or within the class of all S-worlds, plus any one extra world, minus any S-worlds that are M-equivalent to the extra world.

149

To exclude these trivial cases, we need to impose a condition roughly as follows: the class X must contain a majority of the worlds where S is true, and also a majority of the worlds where S is false. Henceforth when we speak of partial supervenience, let us always mean partial supervenience within a class that satisfies this condition of non-triviality.

But what should we mean by 'a majority'? If there were finitely many worlds, we could just count; but there are infinitely many worlds. We could require a difference in cardinality, infinite or otherwise; but that would make the condition altogether too stringent.

Instead, we make the bold conjecture that we are given a certain probability distribution over the worlds, call it 'Prob', which would represent a reasonable initial distribution of subjective probability prior to all experience.[21] Then we may say that the condition is satisfied iff Prob(X/S) and Prob(X/Not-S) both exceed 50%. (This requires that Prob(S) and Prob(Not-S) are positive, else the conditional probabilities would be undefined.)

Here is an example of partial aboutness in the sense of (non-trivial) partial supervenience. Suppose we have an urn with 100 balls, some but not all of them green. The frequency of green balls in the urn is a subject matter. Suppose we sample randomly, with replacement, for very many draws. It is always possible to draw an unrepresentative sample, but with our very large sample it is very improbable. Let X contain all the worlds where the sample is representative: that is, where the sample frequency, rounded to the nearest percent, equals the urn frequency. Let X also contain all worlds contrary to our stipulation of the situation. Sample frequency does not supervene *simpliciter* on urn frequency − you can still get any sample from any urn − but it does supervene on urn frequency within X. So the truth value of a statement S which specifies the sample frequency

21 If we imagine this distribution to be uniquely determined, we have made altogether too bold a conjecture. But let us suppose instead that we have a class of reasonable initial probability distributions, differing somewhat but not too much from one another; and that what follows is said relative to some arbitrary choice from that class. As usual, what is true on all ways of making the arbitrary choice is determinately true; what is false on all ways is determinately false; what is true on some and false on others is indeterminate.

(rounded to the nearest percent) likewise supervenes on urn frequency within X. Our condition of non-triviality is satisfied – very well satisfied, since the overwhelming majority of S-worlds, and the overwhelming majority of (Not-S)-worlds, and also the overwhelming majority of (Not-S)-worlds satisfying our stipulation, all fall within X. So we may say that S is partly about the urn frequency, in the sense of partial supervenience. I do find it fairly natural to say this.

I think we could find it no less natural to say that a statement is partly about observation if it is so in the sense of partial supervenience – at least, if the condition of non-triviality is more than barely satisfied. But that scarcely matters. What does matter is that we have here a third line of retreat from the idea that an empirical statement is entirely about observation, and one that can again be linked to Ayer's discussion. The link, however, is not via the criterion – as we shall see, both of the guiding principles we took from it are violated. Rather, the link is to Ayer's preliminary suggestion, before the criterion, that the test question for a putative statement of fact is: 'Would any observation be relevant to the determination of its truth or falsehood?'

Ayer found the notion of evidential relevance unclear; but for us, with a well-developed probabilistic model of confirmation, it is in good shape. We have assumed that we are given a certain reasonable initial probability distribution, Prob. Then we may say that E is evidentially relevant to S iff Prob(S/E) differs from Prob(S). Iff some statement entirely about a subject matter is evidentially relevant to S, we may say the same about the subject matter itself. Then observation is evidentially relevant to S iff, for some statement 0 entirely about observation, Prob(S/0) differs from Prob(S).

A statement S is partly about observation, in the sense of partial supervenience, iff observation is evidentially relevant to S. Or rather, this is so *modulo* two idealisations; I shall omit a precise statement of the result, and allow the idealisations to appear in the course of the proof.

Proof. Left to right. S supervenes on observation within a class X that satisfies our condition of non-triviality. We can assume without loss of generality that any S-world observationally equivalent to an X-world where S is true is itself in X, and any (Not-S)-world observationally equivalent to an X-world where S is false is itself in X. (For if it were

151

not so originally, we could just add the missing worlds to X, and the new expanded X would satisfy non-triviality as well as the old X did.) Assume, by way of idealisation, that for any class of worlds, there is a statement true at exactly the worlds in that class. Let O be a statement true at any world observationally equivalent to an X-world where S is true. Let P be true at any world observationally equivalent to an X-world where S is false. Let Q be true at any world that is not observationally equivalent to any X-world. These three statements are entirely about observation, and they are mutually exclusive and jointly exhaustive. Suppose for *reductio* that none of them is evidentially relevant to S: Prob(S/O), Prob(S/P), Prob(S/Q) are all equal. Then for X to satisfy our condition of non-triviality, Prob(O) and Prob(P) both must be greater than 50%; which is impossible.

Right to left. Consider the equivalence classes under the relation of observational equivalence. Divide them into 'upper' and 'lower' classes such that, first, the two classes differ as little as possible in total probability, and second, whenever A is in the upper class and B is in the lower, Prob(S/A) is greater than or equal to Prob(S/B). Then also, whenever A is in the upper class and B is in the lower, Prob(Not-S/A) is less than or equal to Prob(Not-S/B). Since observation is evidentially relevant to S, we will sometimes have inequality. Let U be the union of the upper class, and let L be the union of the lower class. Then Prob(S/U) exceeds Prob(S/L), and Prob(Not-S/L) exceeds Prob(Not-S/U). We made Prob(U) and Prob(L) approximately equal; if the approximation is good enough – now we assume, by way of idealisation, that it can be made good enough – it follows that Prob(U/S) and Prob(L/Not-S) both exceed 50%. Let class X contain the worlds in U where S is true together with the worlds in L where S is false. Then S supervenes on observation within X, and X satisfies our condition of non-triviality. *QED*

Given that partial aboutness in the present sense amounts to evidential relevance, it is easy to see how it violates both the Entailment Principle and the Compositional Principle. In fact, it can violate both at once. It can happen that observation is relevant to P, and also to Q, but not to their conjunction 'P & Q'. (And further, that 'P & Q' is not analytic.) Then the Entailment Principle is violated because 'P & Q' entails P, and the Compositional Principle is violated because 'P & Q' is a truth-functional compound of P and Q.

Miniature example. We have just four worlds, all equally probable. We have two observational alternatives: L and D (light and dark; left and right column). P-worlds are drawn as noughts, Q-worlds as crosses, (P & Q)-worlds therefore as noughts superimposed on crosses.

L: $\oplus$ +   D: O $\oplus$
Prob(P/L) = 50%   Prob(P/D) = 100%
Prob(Q/L) = 100%   Prob(Q/D) = 50%
Prob(PQ/L) = 50%   Prob(PQ/D) = 50%

Observation is relevant to P and to Q, but not to 'P & Q'.

Partial aboutness in the sense of partial supervenience – that is, evidential relevance – does obey the Equivalence Principle. The present conception, like the previous ones, has no resources to distinguish between equivalent statements. If two statements are equivalent, they supervene on exactly the same subject matters, within any class; and their evidential relations are the same.

As a delineation of the empirical, being partly about observation in the sense of partial supervenience – that is, evidential relevance of observation – again seems acceptable, though again it is only one candidate among others.[22] But again it is hopeless as a standard of meaningfulness, because it is absurd that we should be able to make a meaningless statement just by conjoining two meaningful ones.

XI

Fourth, we have the *part-of-statement* conception: a statement is partly about a subject matter iff some part of that statement is entirely about that subject matter. This presupposes that statements have other statements as parts. Do they? No, if we conceive of statements as propositions, and propositions just as sets of possible worlds.* Yes,

---

22 See Brian Skyrms, *Pragmatics and Empiricism* (New Haven and London: Yale University Press, 1984) 14–19 and 111–119, for further discussion of evidential relevance to observation, understood in terms of probability, as a way of delineating the empirical.

* [Added 1996] I cannot think why I said this: I already believed that sets had their own subsets as parts. (See 'Mathematics is Megethology' in this volume.)

if we conceive of statements as 'structured meanings', abstracted from sentences far enough to leave behind such superficial details as the spelling and pronunciation and order of words, but not far enough to leave behind the syntactic structure which divides a sentence into constituent clauses.[23] I believe that these conceptions (and others) are entirely legitimate. There is no saying which one better deserves the name 'statement', and no saying which one better fits what Ayer had in mind.

Consider these three sentences, equivalent but different with respect to their subsentences.

(a) The Absolute is crafty.
(b) Either the Absolute is crafty and it's dark, or else the Absolute is crafty and it isn't dark.
(c) The Absolute is crafty, and either it's dark or it isn't dark.

We can perfectly well say that we have the same statement, expressed three different ways. Or we can just as well say that we have three different, but equivalent, statements. In the second case, we will say that just as the sentences (a)–(c) have different sentences as parts, so likewise the corresponding statements (a)–(c) have different statements as parts. That gives us what we need to make sense of the part-of-statement conception of partial aboutness.

Our previous conceptions of entire and partial aboutness were all *intensional:* if there ever were two nonidentical equivalent statements, they wouldn't differ in aboutness. So we never had to choose between conceptions of statements that do or don't allow nonidentical equivalence. The part-of-statement conception, on the other hand, is *hyperintensional:* it distinguishes between equivalent statements. The statements (b) and (c) are partly about observation, but the statement (a) is not. The reason is that (b) and (c), unlike (a), have as a part the statement 'It's dark', which is entirely about obser-

---

23 See M.J. Cresswell, *Structured Meanings: The Semantics of Propositional Attitudes* (Cambridge, Massachusetts: M.I.T. Press, 1985); or the discussion of 'meanings' in my 'General Semantics', *Synthese* 22 (1970) 18–67.

vation. Also, if we look at parts that are already compound, we find that (c), unlike (a) and (b), is partly about tax reform, or any subject matter whatever. The reason is that (c), unlike (a) and (b), has as a part the analytic statement 'Either it's dark or it isn't', and an analytic statement is entirely about anything.

The part-of-statements conception is *cumulative*. When we build up statements from their parts, we may gain new subject matters for the resulting statement to be partly about, but we never lose old ones. Not so for our other conceptions, even applied under the assumption that statements have other statements as parts. On our other conceptions, (b) and (c) would not be partly about observation, despite the presence within them of a statement which is entirely about observation.

The part-of-statements conception deserves attention for the sake of completeness, and because close relatives of it are prominent in other discussions.[24] However, I see no way of linking it with what Ayer said. Neither the evidential relations of Ayer's preliminary suggestion nor the entailment relations of the criterion itself are sensitive to hyperintensional distinctions. And if we are seeking something that can pass for a delineation of the empirical, we scarcely want something that will admit (b) and (c) while excluding (a).

XII

When something goes bump in the night, it's none too reassuring to be told there's nothing there. You'll sleep more soundly when you know there is something there, but only Magpie and Possum. When Ayer threatened us with the criterion, the collapse tried to tell us there was nothing there. Unconvinced, the patch-and-puncture industry struggles on. Well, there was something there. Or rather, several things – but no fear, nothing that could possibly carry us off to old Vienna. Now perhaps we can rest.

---

24 Two treatments of hyperintensional and cumulative conceptions of aboutness, or 'relevance to a context', are Richard L. Epstein, 'Relatedness and Implication', *Philosophical Studies* 36 (1979) 137–173 (see especially 156–158); and B.J. Copeland, 'Horseshoe, Hook, and Relevance,' *Theoria* 50 (1984) 148–164. In both papers, however, the aboutness of atomic statements is left unanalysed.

10

Ayer's first empiricist criterion of meaning: why does it fail?

In the first edition of *Language, Truth and Logic*, Ayer proposes that 'the mark of a genuine factual proposition' is 'that some experiential propositions can be deduced from it in conjunction with certain other premises without being deducible from those other premises alone' ([1], pp. 38–9).

Berlin objects that the criterion is 'a good deal too liberal' and admits patent nonsense. His example, near enough, is that the experiential proposition 'I dislike this logical problem' can be deduced from the nonsensical 'this logical problem is green' in conjunction with the premise 'I dislike whatever is green' without being deducible from the latter premise alone ([2], p. 234). Berlin's point is well taken, if indeed a category mistake is nonsense. But it shows at most that the criterion admits too much. We do not yet know how much too much.

In his introduction to the second edition, Ayer accepts Berlin's point in a greatly extended form. Not only does the criterion admit this or that piece of nonsense; it 'allows meaning to any statement whatsoever. For, given any statement "S" and an observation-statement "O", "O" follows from "S" and "if S then O" without following from "if S then O" alone' ([1], p. 11).

Here Ayer goes wrong. For it may very well happen that the consequent of a conditional 'if S then O' does follow from that conditional alone. For instance, O follows from 'if (P or not P) then O'

First published in *Analysis* 48 (1988), 1–3.

and from 'if not O, then O'. Ayer has just overlooked such cases. In general, O follows from 'if S then O' just when 'S or O' is analytic.

(Two remarks. (1) I assume that Ayer's 'if . . . then' is meant truth-functionally. (2) I assume that 'follows' and 'analytic' are to be taken in corresponding senses, so that a conclusion follows from a premise just when the conditional from premise to conclusion is analytic. Then if 'analytic' covers more than narrowly logical truth, 'follows' must likewise cover more than narrowly logical implication. Alternatively, 'follows' can be glossed as 'follows by logic alone' provided that 'analytic' is glossed as 'logically true'. My point goes through either way.)

It should have been easy to see that the criterion does not admit just any statement as factual. For if S is analytic, nothing at all follows from S in conjunction with other premises, except for what follows from those other premises alone. To be sure, Ayer does count analytic statements as meaningful. But they are definitely not supposed to come out as factual; rather, the meaningful divides into the analytic and the factual ([1], p. 31).

Ayer is right, however, that the criterion 'allows meaning to any statement whatsoever'. Only his proof is at fault, not its conclusion. Any statement S is either factual by the criterion or else analytic, and counts as meaningful either way. The correct proof is by cases. First case: for some observation-statement O, 'S or O' is not analytic. Then O does not follow from 'if S then O' alone, but does follow from S in conjunction with 'if S then O'. Then S is factual by the criterion. Second case: there is no such O. It is safe to assume that some two observation-statements O_1 and O_2 conflict, so that 'not both O_1 and O_2' is analytic. Also 'S or O_1' and 'S or O_2' are analytic *ex hypothesi*, and from these three analytic premises S follows. Then S itself is analytic. Q.E.D. Ayer's mistake was simply to omit the second case.

Later authors have perpetuated the mistake. Church writes that in Ayer's second edition, 'A criticism by Berlin . . . is accepted as valid, according to which the definition as actually given in the first edition would make all statements verifiable' ([3], p. 52). Church himself implicitly endorses the criticism thus stated; yet it is mistaken, since Ayer uses 'verifiable' interchangeably with 'factual'. Further, Berlin said nothing so general. Several others paraphrase and endorse Ayer's faulty argument and conclusion: Hempel ([5], p. 49), Scheffler ([6], p. 41), and recently Foster ([4], p. 14).

REFERENCES

[1] A. J. Ayer, *Language, Truth and Logic,* Gollancz, 1936; second edition, 1946. Page references are to the second edition.
[2] Isaiah Berlin, 'Verification', *Proceedings of the Aristotelian Society,* 39 (1938–39).
[3] Alonzo Church, review of the second edition of *Language, Truth and Logic, Journal of Symbolic Logic,* 14 (1949).
[4] John Foster, *Ayer,* Routledge and Kegan Paul, 1985.
[5] Carl G. Hempel, 'Problems and Changes in the Empiricist Criterion of Meaning', *Revue Internationale de Philosophie,* 4 (1950).
[6] Israel Scheffler, *The Anatomy of Inquiry,* Knopf, 1963.

11

Analog and digital

The distinction between analog and digital representation of numbers is well understood in practice. Yet its analysis has proved troublesome. I shall first consider the account given by Nelson Goodman and offer examples to show that some cases of analog representation are mis-classified, on Goodman's account, as digital. Then I shall offer alternative analyses of analog and digital representation.

1. DIFFERENTIATED ANALOG REPRESENTATION

According to Goodman in *Languages of Art*,[1] the distinction between digital and analog representation of numbers is as follows. Digital representation is *differentiated*. Given a number-representing "mark" – an inscription, vocal utterance, pointer position, electrical pulse, or whatever – it is theoretically possible, despite our inability to make infinitely precise measurements, to determine exactly which other marks are copies of the given mark and to determine exactly which number (or numbers) the given mark and its copies represent. Analog representation, on the other hand, fails to be differentiated because it is *dense*. For any two marks that are not copies, no matter

First published in *Noûs* 5 (1971), 321–327. Reprinted with kind permission from Blackwell publishers.

1 (Indianapolis and New York: Bobbs-Merrill, 1968), sections IV.2, IV.5, and IV.8. I have combined Goodman's syntactic and semantic differentiation and combined his syntactic and semantic density; and I have not defined differentiation and density in full generality, but only as applied to the representation of numbers.

how nearly indistinguishable they are, there could be a mark intermediate between them which is a copy of neither; and for any two marks that are not copies and represent different numbers, no matter how close the numbers are, there is an intermediate number which would be represented by a mark that is a copy of neither.

It is true and important that digital representation is differentiated, and that it differs thereby from the many cases of analog representation that are undifferentiated and dense: those cases in which all real numbers in some range are represented by values of some continuously variable physical magnitude such as voltage. But there are other cases: representation that is as differentiated and non-dense as any digital representation and yet is analog rather than digital representation. Here are two examples of differentiated analog representation. If accepted, they show that Goodman's distinction, interesting though it is in its own right, does not coincide with the analog-digital distinction of ordinary technological language.

Example 1: ordinary electrical analog computers sometimes receive their numerical inputs in the form of settings of variable resistors. A setting of 137 ohms represents the number 137, and so on. There are two ways these variable resistors might work. In the first case, a contact slides smoothly along a wire with constant resistance per unit length. In the second case, there is a switch with a very large but finite number of positions, and at each position a certain number of, say, 1-ohm fixed resistors are in the circuit and the rest are bypassed. I do not know which sort of variable resistor is used in practice. In either case, the computer is an analog computer and its representation of numbers by electrical resistances is analog representation. In the sliding-contact case the representation is undifferentiated and dense; but in the multi-position switch case the representation is differentiated and non-dense, yet analog and not digital.

Example 2: we might build a device which works in the following way to multiply two numbers x and y. There are four receptacles: W, X, Y, and Z. We put a large amount of some sort of fluid – liquid, powder, or little pellets – in W. We put x grams of fluid in X, y grams in Y, and none in Z. At the bottom of X is a valve, allowing fluid to drain at a constant rate from X into a wastebasket. At the bottom of W is a spring-loaded valve, allowing fluid to drain from W into Z at a rate proportional to the amount of fluid in Y. For instance, if Y contains 17 grams of fluid, then the rate of drainage from W is 17 times the constant rate of drainage from X. We simultaneously open the valves on W and X; as soon as X is empty, we close the valve on W. The amount z of fluid that has passed from W to Z at a rate proportional to y in a time proportional to x is the product of the numbers x and y. This device is an analog computer, and its representation of numbers by amounts of fluid is analog representation. In case the fluid is a liquid, the representation is undifferentiated and dense (almost – but even a liquid consists of molecules); but in case it is 1-gram round metal pellets, the representation is differentiated and non-dense, yet analog and not digital.

2. ANALOG REPRESENTATION

It is commonplace to say that analog representation is representation of numbers by physical magnitudes. And so it is; but so is digital representation, or any other sort of representation that could be used in any physically realized computer, unless we adopt a peculiarly narrow conception of physical magnitudes.

We may regard a physical magnitude as a function which assigns numbers to physical systems at times. A physical magnitude may be defined for every physical system at every moment of its existence, or it may be defined only for physical systems of some particular kind or only at some times. Let us call any such function a *magnitude*. We can then define a *physical magnitude* as any magnitude definable in the language of physics more or less as we know it. This is imprecise, but properly so: the vagueness of "physical magnitude" ought to correspond to the vagueness of "physics".

Since the language of physics includes a rich arithmetical vocabulary, it follows that if the values of a magnitude depend by ordinary arithmetical operations on the values of physical magnitudes, then that magnitude is itself a physical magnitude. Now suppose that in computers of a certain sort, numbers are represented in the following way. We consider the voltages $v_0, \ldots, v_{35}$ between 36 specified pairs of wires in any such computer at any time at which the computer is operating. By taking these voltages in order, and associating with each positive voltage the digit 1 and with each negative voltage the digit 0, we obtain a binary numeral and the number denoted by that numeral. Here is as good a case of digital representation as we could find. Yet it is also a case of representation by a physical magnitude: the number represented thus by computer s at time t is the value of the physical magnitude defined below by means of arithmetical vocabulary in terms of the voltages $v_0, \ldots, v_{35}$. (We take the magnitude to be undefined if any v_i is 0.)

$$V(s,t) = \sum_{i=0}^{35} 2^i \begin{Bmatrix} 1 \text{ if } v_i(s,t) > 0 \\ 0 \text{ if } v_i(s,t) < 0 \end{Bmatrix}$$

$$= \sum_{i=0}^{35} 2^i \left[\frac{1}{2} + \frac{\sqrt{v_i(s,t)^2}}{2v_i(s,t)} \right]$$

Analog representation, then, is representation of numbers by physical magnitudes of a special kind. Resistances, voltages, amounts of fluid, for instance, are physical magnitudes of the proper kind for analog representation; but V, as defined above, is a physical magnitude not of the proper kind for analog representation. How may we distinguish physical magnitudes that are of the proper kind for analog representation from those that are not?

We might try saying that the magnitudes suitable for analog representation are those that are expressed by primitive terms in the language of physics. This will not do as it stands: a term is primitive not relative to a language but relative to some chosen definitional reconstruction thereof. Any physical magnitude could be expressed by a primitive term in a reconstruction designed *ad hoc,* and there is no physical magnitude that must be expressed by a primitive term. We may, however, define a *primitive magnitude* as any physical magnitude that is expressed by a primitive term in some *good* reconstruction of the language of physics – good according to our ordinary standards of economy, elegance, convenience, familiarity. This definition is scarcely precise, but further precision calls for a better general understanding of our standards of goodness for definitional reconstructions, not for more work on the topic at hand.

Even taking the definition of primitive magnitudes as understood, however, it is not quite right to say that analog representation is representation by primitive magnitudes. Sometimes it is representation by physical magnitudes that are *almost primitive:* definable in some simple way, with little use of arithmetical operations, in terms of one or a few primitive magnitudes. (Further precision here awaits a better general understanding of simplicity of definitions.) Products of a current and a voltage between the same two points in a circuit, or logarithmically scaled luminosities, seem unlikely to be expressed by primitive terms in any good reconstruction. But they are almost primitive, and representation of numbers by them would be analog representation. Of more interest for our present purposes, rounded-off primitive magnitudes are almost primitive. Such are the number-representing magnitudes in our two examples of differentiated analog representation: resistance rounded to the nearest ohm and amount of fluid rounded to the nearest gram. We use the rounded-off magnitudes because, in practice, our resistors or pellets are not exactly one ohm or gram each; so if we take the resistance of 17 resistors or the mass of 17 pellets to represent the number 17 (rather than some unknown number close to 17) we are using not primitive magnitudes but almost primitive magnitudes.

The commonplace definition of analog representation as representation by physical magnitudes is correct, so far as I can see, if taken as follows: *analog representation* of numbers is representation of numbers by physical magnitudes that are either primitive or almost primitive according to the definitions above.

3. DIGITAL REPRESENTATION

It remains to analyze digital representation, for not all non-analog representation of numbers is digital. (It may be, however, that all practically useful representation of numbers is either analog or digital.) If numbers were represented by the physical magnitude defined as follows in terms of voltages between specified pairs of wires in some circuit, the representation would be neither analog nor digital (nor useful).

$$P(s,t) = \sum_{i=0}^{35} \log_{10} (\sinh(\sqrt{v_i(s,t)}))$$

Digital representation is representation by physical magnitudes of another special kind, the kind exemplified above by V.

We may first define an *n-valued unidigital magnitude* as a physical magnitude having as values the numbers $0, 1, \ldots, n-1$ whose values depend by a step function on the values of some primitive magnitude. Let U be an n-valued unidigital magnitude; then there is a primitive magnitude B which we may call the *basis* of U and there is an increasing sequence of numbers $a_1, \ldots, a_{n-1}$ which we may call the *transition points* of U and, for each system s on which U is defined, there is a part p of s such that U is defined as follows on s.

$$U(s,t) = \left\{ \begin{array}{l} 0 \text{ if } a_1 > B(p,t) \\ 1 \text{ if } a_2 > B(p,t) > a_1 \\ \quad \cdot \\ \quad \cdot \\ \quad \cdot \\ n-1 \text{ if } B(p,t) > a_{n-1} \end{array} \right\}$$

$$= \sum_{i=1}^{n-1} \left[\frac{1}{2} + \frac{\sqrt{(B(p,t) - a_i)^2}}{2(B(p,t) - a_i)} \right]$$

Let us call a unidigital magnitude *differentiated* if, in the systems on which it is defined, its basis does not take values at or near its transition points.

Representation of numbers by differentiated unidigital magnitudes, or by physical magnitudes whose values depend arithmetically on the values of one or more differentiated unidigital magnitudes, is differentiated and non-dense representation, hence digital representation in Goodman's sense. But it may not really be digital representation. In fact, it may be analog representation. A unidigital magnitude with many evenly spaced transition points is exactly what we have been calling a rounded-off primitive magnitude; it is almost primitive and suitable for analog representation. The number-representing magnitudes in our examples of differentiated analog representation are differentiated unidigital magnitudes, but representation by these magnitudes is analog representation.

What distinguishes digital representation, properly so-called, is not

164

merely the use of differentiated unidigital magnitudes; it is the use of the many combinations of values of a few few-valued unidigital magnitudes. Let us now define a *multidigital magnitude* as any physical magnitude whose values depend arithmetically on the values of several few-valued unidigital magnitudes. Let us call a multidigital magnitude *differentiated* if it depends on differentiated unidigital magnitudes. In fixed point digital representation, for instance, a multidigital magnitude M depends as follows on several n-valued unidigital magnitudes $u_0, \ldots, u_{m-1}$.

$$M(s,t) = \sum_{i=0}^{m-1} n^i \, u_i(s,t)$$

V is a multidigital magnitude of this sort, with $m = 36$, $n = 2$, and each u_i having as its basis the corresponding voltage v_i at the specified part of the system. Other multidigital magnitudes depend in more complicated ways upon their unidigital magnitudes. Most often, the unidigital magnitudes are not merely few-valued but two-valued; but not so, for instance, in an odometer in which the unidigital magnitudes are ten-valued step functions of angles of rotation of the wheels.

I suggest, therefore, that we can define *digital representation* of numbers as representation of numbers by differentiated multidigital magnitudes.

12

Lucas against mechanism

J. R. Lucas argues in "Minds, Machines, and Gödel"[1] that his potential output of truths of arithmetic cannot be duplicated by any
Turing machine, and *a fortiori* cannot be duplicated by any machine.
Given any Turing machine that generates a sequence of truths of
arithmetic, Lucas can produce as true some sentence of arithmetic
that the machine will never generate. Therefore Lucas is no machine.

I believe Lucas's critics have missed something true and important
in his argument. I shall restate the argument in order to show this.
Then I shall try to show how we may avoid the anti-mechanistic
conclusion of the restated argument.

As I read Lucas, he is rightly defending the soundness of a certain
infinitary rule of inference. Let L be some adequate formalization of
the language of arithmetic; henceforth when I speak of sentences, I
mean sentences of L, and when I call them true, I mean that they are
true on the standard interpretation of L. We can define a certain
effective function *Con* from machine tables to sentences, such that
we can prove the following by metalinguistic reasoning about L.

C1. Whenever M specifies a machine whose potential output is a set
    S of sentences, *Con* (M) is true if and only if S is consistent.

First published in *Philosophy* 44 (1969), 231–233. Reprinted with kind permission
from *Philosophy*.

I am indebted to George Boolos and Wilfrid Hodges for valuable criticisms of an
earlier version of this paper.

1 *Philosophy*, 36 (1961): 112–127.

C2. Whenever M specifies a machine whose potential output is a set S of true sentences, *Con* (M) is true.

C3. Whenever M specifies a machine whose potential output is a set S of sentences including the Peano axioms, *Con* (M) is provable from S only if S is inconsistent.

Indeed, there are many such functions; let *Con* be any chosen one of them. Call Φ a *consistency sentence* for S if and only if there is some machine table M such that Φ is *Con* (M) and S is the potential output of the machine whose table is M. Now I can state the rule R which I take Lucas to be defending.

R. If S is a set of sentences and Φ is a consistency sentence for S, infer Φ from S.

Lucas's rule R is a perfectly sound rule of inference: if the premises S are all true, then by C2 so is the conclusion Φ. To use R is to perform an inference in L, not to ascend to metalinguistic reasoning about L. (It takes metalinguistic reasoning to show that R is truth-preserving, but it takes metalinguistic reasoning to show that *any* rule is truth-preserving.)

Lucas, like the rest of us, begins by accepting the Peano axioms for arithmetic. (Elementary or higher-order; it will make no difference.) A sentence Ψ is a theorem of Peano arithmetic if and only if Ψ belongs to every superset of the axioms which is closed under the ordinary rules of logical inference. Likewise, let us say that a sentence χ is a theorem of *Lucas arithmetic* if and only if χ belongs to every superset of the axioms which is closed under the ordinary rules of logical inference and also closed under Lucas's rule R. We have every bit as much reason to believe that the theorems of Lucas arithmetic are true as we have to believe that the theorems of Peano arithmetic are true: we believe the Peano axioms, and the theorems come from them by demonstrably truth-preserving rules of inference. Knowing this, Lucas stands ready to produce as true any theorem of Lucas arithmetic.[2]

2 Lucas arithmetic belongs to a class of extensions of Peano arithmetic studied by A. M. Turing in "Systems of Logic Based on Ordinals", *Proceedings of the London*

Suppose Lucas arithmetic were the potential output of some Turing machine. Then it would have a consistency sentence Φ. Since Lucas arithmetic is closed under R, Φ would be a theorem of Lucas arithmetic. Then Φ would, trivially, be provable from Lucas arithmetic. Then, by C3, Lucas arithmetic would be inconsistent. Lucas arithmetic would contain falsehoods, and so would the Peano axioms themselves. Therefore, insofar as we trust the Peano axioms, we know that Lucas arithmetic is not the potential output of any Turing machine. Assuming that any machine can be simulated by a Turing machine – an assumption that can best be taken as a partial explication of Lucas's concept of a machine – we know that neither is it the potential output of any machine. Thus if Lucas arithmetic is the potential output of Lucas, then Lucas is no machine.

So far, so good; but there is one more step. Although Lucas has good reason to believe that all the theorems of Lucas arithmetic are true, it does not yet follow that his potential output is the whole of Lucas arithmetic. He can produce as true any sentence which he can somehow *verify* to be a theorem of Lucas arithmetic. If there are theorems of Lucas arithmetic that Lucas cannot verify to be such, then his potential output falls short of Lucas arithmetic. For all we know, it might be the potential output of a suitable machine. To complete his argument that he is no machine – at least, as I have restated the argument – Lucas must convince us that he has the necessary general ability to verify theoremhood in Lucas arithmetic. If he has that remarkable ability, then he can beat the steam drill – and no wonder. But we are given no reason to think that he does have it.

It is no use appealing to the fact that we can always verify theoremhood in any ordinary axiomatic theory – say, Peano arithmetic – by exhibiting a proof. True, if we waive practical limitations on endurance; but Lucas arithmetic is not like an ordinary axiomatic theory. Its theorems do have proofs; but some of these proofs are transfinite sequences of sentences since Lucas's rule R can take an infinite set S of premises. These transfinite proofs will not be discov-

Mathematical Society, sec. 2, 45 (1939): 161–228, and by S. Feferman in "Transfinite Recursive Progressions of Axiomatic Theories", *Journal of Symbolic Logic,* 27 (1962): 259–316.

ered by any finite search, and they cannot be exhibited and checked in any ordinary way. Even the finite proofs in Lucas arithmetic cannot be checked by any mechanical procedure, as proofs in an ordinary axiomatic theory can be. In order to check whether Lucas's rule R has been used correctly, a checking procedure would have to decide whether a given finite set S of sentences was the output of a machine with a given table M. But a general method for deciding that could easily be converted into a general method for deciding whether any given Turing machine will halt on any given input – and that, we know, is impossible.

We do not know how Lucas verifies theoremhood in Lucas arithmetic, so we do not know how many of its theorems he can produce as true. He can certainly go beyond Peano arithmetic, and he is perfectly justified in claiming the right to do so. But he can go beyond Peano arithmetic and still be a machine, provided that some sort of limitations on his ability to verify theoremhood eventually leave him unable to recognize some theorem of Lucas arithmetic, and hence unwarranted in producing it as true.

13

Lucas against mechanism II

J. R. Lucas serves warning that he stands ready to refute any sufficiently specific accusation that he is a machine. Let any mechanist
say, to his face, that he is some particular machine M; Lucas will respond by producing forthwith a suitable Gödel sentence ϕ_M. Having
produced ϕ_M, he will then argue that – given certain credible
premises about himself – he could not have done so if the accusation
that he was M had been true. Let the mechanist try again; Lucas will
counter him again in the same way. It is not possible to accuse Lucas
truly of being a machine.[1]

I used to think that the accusing mechanist interlocutor was an
expository frill, and that Lucas was really claiming to be able to do
something that no machine could do.[2] But I was wrong; Lucas insists
that the interlocutor does play an essential role. He writes that "the
argument is a dialectical one. It is not a direct proof that the mind is
something more than a machine; but a schema of disproof for any
particular version of mechanism that may be put forward. *If* the
mechanist maintains any specific thesis, I show that a contradiction
ensues. But only if. It depends on the mechanist making the first
move and putting forward his claim for inspection."[3] Very well. I

First published in *The Canadian Journal of Philosophy* 9 (1979), 373–376. Reprinted
with kind permission from *The Canadian Journal of Philosophy*.

1 J. R. Lucas, "Minds, Machines and Gödel," *Philosophy* 36 (1961), pp. 112–27.
2 David Lewis, "Lucas Against Mechanism," *Philosophy* 44 (1969), pp. 231–33;
   reprinted as Chapter 12 of this volume.
3 J. R. Lucas, "Satan Stultified: A Rejoinder to Paul Benacerraf," *Monist* 52 (1968),

promise to take the dialectical character of Lucas's argument more seriously this time – and that shall be his downfall.

Let O_L be Lucas's potential arithmetical output (i.e., the set of sentences in the language of first order arithmetic that he is prepared to produce) when he is not accused of being any particular machine; and for any machine M, let O_L^M be Lucas's arithmetical output when accused of being M. Lucas himself has insisted (in the passage I quoted) that the mechanist's accusations make a difference to his output. Therefore we cannot speak simply of Lucas's arithmetical output, but must take care to distinguish O_L from the various O_L^M's.

Likewise for any machine M: let O_M be M's arithmetical output when not accused of being any particular machine, and let O_M^N be M's arithmetical output when accused of being some particular machine N. If the machine M, like Lucas, is capable of responding to accusations, then O_M and the various O_M^N's may differ.

We may grant Lucas three premises.

(1) (Every sentence of) O_L is true. For O_L is nothing else but everyman's arithmetical lore, and to doubt the truth thereof would be extravagant scepticism.

(2) O_L includes all the axioms of Elementary Peano Arithmetic. Lucas can easily convince us of this.

(3) For any machine M, O_L^M consists of O_L plus the further sentence ϕ_M, a Gödel sentence expressing the consistency of M's arithmetical output. It is Lucas's declared policy thus to respond to any mechanistic accusation by producing the appropriate Gödel sentence; and – ignoring, for the sake of the argument, any practical limits on Lucas's powers of computation – he is able to carry out this plan. (We may take it that a mechanistic accusation is not sufficiently specific to deserve refutation unless it provides Lucas with a full functional specification of the machine he is accused of being: a machine table or the like.)

pp. 145–46. See also J. R. Lucas, "Mechanism: A Rejoinder," *Philosophy* 45 (1970), pp. 149–51; and J. R. Lucas, *The Freedom of the Will* (Oxford, 1970), pp. 139–45.

Let the mechanist accuse Lucas of being a certain particular machine M. Suppose by way of *reductio* that the accusation is true. Then $O_L = O_M$ and $O_L^M = O_M^M$.

M is a machine. In the present context, to be a machine is not to be made of cogwheels or circuit chips, but rather to be something whose output, for any fixed input, is recursively enumerable. (More precisely, the set of Gödel numbers encoding items of output is recursively enumerable.) If the whole output of M, on input consisting of a certain mechanistic accusation, is recursively enumerable, then so is the part that consists of sentences of arithmetic: O_M^M, in the case under consideration.

Then there is an axiomatizable formal theory θ that has as theorems all and only the sentences of arithmetic that are deducible in first order logic from O_M^M. Further, θ is an extension of Elementary Peano Arithmetic: by premise (2) the axioms thereof belong to O_L, by premise (3) O_L is included in O_L^M, O_L^M – that is, O_M^M – is included in θ. Hence θ is the sort of theory that cannot contain a Gödel sentence expressing its own consistency unless it is inconsistent.

Is θ inconsistent? Apparently so. The Gödel sentence ϕ_M belongs to O_L^M, hence to O_M^M, and hence to θ.

Yet if ϕ_M is true, then O_L^M, which is O_L plus ϕ_M, is true by premise (1); hence O_M^M is true, hence θ is true and *a fortiori* consistent.

Lucas says that he can see that ϕ_M *is* true. Surely he means that he can see that *if* the accusation that he is M is true, *then* ϕ_M is true. If he meant more than that, the accusation – which he disbelieves and is in process of refuting – is irrelevant; he ought to be able to see that ϕ_M is true without the accuser's aid, contrary to his insistence on the dialectical character of his argument.

How could he see that? Perhaps as follows. (I can see no other way.) By premise (1), Lucas's arithmetical output is true. If true, then *a fortiori* it is consistent. If the accusation that Lucas is M is true, it follows that the arithmetical output of M is consistent. Accordingly, a Gödel sentence expressing the consistency thereof is true – and ϕ_M is just such a sentence.

And so the supposition that Lucas is M has seemingly led to contradiction. On the one hand, θ contains ϕ_M and must therefore be inconsistent; on the other hand ϕ_M is true, so θ is true, so

θ is consistent. The mechanistic accusation stands refuted. Q.E.D.

Not quite! We must be more careful in saying what ϕ_M is. It is, we said, "a Gödel sentence expressing the consistency of M's arithmetical output". Does ϕ_M then express the consistency of O_M, M's arithmetical output when not accused of being any machine? Or of O_M^M, M's arithmetical output when accused of being M? After all, under the supposition that Lucas is M, M has in fact been accused of being M and M's arithmetical output may well have been modified thereby.

First case: ϕ_M is a Gödel sentence expressing the consistency of O_M, M's original arithmetical output unmodified by any accusation. Then we have a correct proof (given premise (1)) that if Lucas is M, then ϕ_M is true. But this ϕ_M does not express the consistency of O_M^M, so it may belong to θ although θ is true and hence consistent. In this case Lucas's *reductio* against the accusation that he is M fails.

Second case: ϕ_M is a Gödel sentence expressing the consistency of O_M^M, M's arithmetical output when accused of being M. Then, since ϕ_M also expresses the consistency of θ, ϕ_M cannot belong to θ unless θ is inconsistent and ϕ_M is therefore false. If Lucas is M, ϕ_M does belong to θ and is false. But so be it. In this case we have no good argument that ϕ_M is true. Even if Lucas is M, ϕ_M no longer expresses the consistency of the trustworthy O_L, but rather of O_L^M: that is, of O_L plus ϕ_M itself. If we tried to argue that ϕ_M is true (if Lucas is M) because it expresses the consistency of a set of truths, we would have to assume what is to be proved: the truth, *inter alia*, of ϕ_M. In this case also Lucas's *reductio* fails.

There are machines that respond to true mechanistic accusations by producing true Gödel sentences of the sort considered in the first case; for all we know, Lucas may be one of them. There are other machines that respond to true mechanistic accusations by producing false Gödel sentences of the sort considered in the second case; for all we know, Lucas may be one of them. Perhaps there also are non-machines, and for all we know Lucas may be one of them.

To confuse the two sorts of Gödel sentences is a mistake. It is part of the mistake of forgetting that the output of Lucas, or of a machine, may depend on the input. And that is the very mistake that Lucas has warned us against in insisting that we heed the dialectical character of his refutation of mechanism.

14

Policing the Aufbau

INTRODUCTION

Carnap's *Aufbau*[1] sketches a remarkably ambitious construction. Given just one primitive phenomenal relation, he seeks to define enough concepts to provide a language adequate for all of science.

The constructed concepts are supposed to be coextensive with certain familiar ones. The *Aufbau* is commonly dismissed as a failure because discrepancies would appear under unfavorable circumstances. That verdict is premature. If there are few discrepancies under actual circumstances, the constructed concepts might be just as adequate for science as the familiar ones they approximate and replace. A mere chance of discrepancies is too bad, but not fatal. It would take frequent discrepancies to spoil the construction, by Carnap's own standards of success. The frequency has not been much investigated – understandably so; the needed computing power has become available only lately. Therefore it remains an open question whether the *Aufbau* succeeds or fails on its own terms.

If the original construction gives too many discrepancies, perhaps a more elaborate version would work better. The version presented,

First published in *Philosophical Studies* 20 (1969), 13–17.

1 Rudolf Carnap, *Der Logische Aufbau der Welt* (Berlin and Schlachtensee: Weltkreis-Verlag, 1928). English translation by Rolf A. George, *The Logical Structure of the World* (Berkeley and Los Angeles: University of California Press, 1967). See also Nelson Goodman's exposition and criticism of the *Aufbau* in *The Structure of Appearance* (Cambridge, Mass.: Harvard University Press, 1951), Chapter V.

after all, is tentative; indeed it serves mostly as an illustration for Carnap's general discussion of logical constructions. There is plenty of room for improvement.

In particular, there are unexploited opportunities to police the construction. Spurious instances of a constructed concept, instances which do not fall under the familiar concept being approximated, often turn out to behave strangely later in the construction. Thereby they may be recognized as suspect and removed. I shall show how this tactic might be used in the early stages of the construction to help fight the difficulty Goodman calls "imperfect community."

QUALITY CLASSES AND THEIR SIMILARITY

Carnap begins his construction with a primitive relation of part similarity (and temporal precedence) between *elementary experiences,* momentary slices of one's total stream of experience (sections 108–10). To speak outside the system, as we must to explain its primitive, two elementary experiences are *part similar* just in case some constituent in one resembles some constituent in the other. There might be a crimson spot at the center of the visual field in one and a scarlet spot slightly off-center in the other, or a smell of skunk in one and a smell of burning rubber in the other, or fright in one and an edgy feeling in the other.

A *similarity circle* (of part similarity) is any maximal class of elementary experiences connected pairwise by part similarity, so that each is part similar to each (section 111). To speak outside the system, a similarity circle is supposed to contain just those elementary experiences in which there occur one or more members of some maximal class of mutually similar constituents. One similarity circle might contain just those elementary experiences in which there occur color spots within a certain small area of the visual field and within certain small ranges of hue, brilliance, and saturation; or just those in which there occur tones within certain small ranges of pitch, loudness, and timbre; or just those in which there occur fright, edgy feelings, or other like emotions.

Unfortunately, not all possible similarity circles are of this intended sort. Spurious similarity circles may appear, connected by a fortuitously complete set of part similarities in miscellaneous respects. Similar centrally located greenish spots may occur in elementary experiences E_1 and E_2, similar lower left brownish spots in E_2 and E_3, similar skunk smells in E_3 and E_4, similar cello tones in E_4 and E_1, similar frights in E_1 and E_3, and similar flute tones in E_2 and E_4. This coincidence ought not to unite E_1, E_2, E_3, and E_4 in a similarity circle, yet it does. Here is the difficulty of imperfect community.

A *quality class* is any maximal subclass of a similarity circle which is wholly included in every similarity circle containing at least half its members (section 112).[2] To speak outside the system, a quality class is supposed to contain just those elementary experiences in which some one constituent occurs. One quality class might contain just those in which crimson of a certain hue, brilliance, and saturation appears at a certain place in the visual field. Hence a similarity circle is supposed to be a union of the quality classes which correspond to several mutually similar constituents. Quality classes are the classes carved out by intersecting similarity circles (ignoring the relatively small intersections due to overlap of the quality classes themselves), and they have been defined accordingly.

Within the system, quality classes could not have been defined in terms of the constituents occurring in elementary experiences; no such things are recognized to exist. Rather the quality classes themselves take the place of constituents; for this reason they are called *quasi constituents* of the elementary experiences, and their construction is called a *quasi analysis.*

Spurious similarity circles will usually yield spurious quality classes: classes of elementary experiences which, as we say outside the system, share no common constituent. We shall see how to distinguish these spurious quality classes from genuine ones by means available within the system itself.

Two quality classes are *similar* just in case each elementary experience in one is part similar to every elementary experience in the

---

2 This is not quite Carnap's definition. I have made the correction proposed by Goodman in *Structure,* Chapter V, section 5. I have also required every quality class to be included in some similarity circle, since otherwise certain unions of intersections of overlapping quality classes would satisfy the definition.

other; that is to say, just in case both are included in some similarity circle (section 114). Speaking outside the system, quality classes are supposed to be similar just in case they correspond to similar constituents. But not all possible similarity is of this intended sort. Two quality classes may be spuriously similar by both being included in some spurious similarity circle.

A *sense class* is any maximal class of quality classes which is connected throughout by chains of similarities (section 115). There is supposed to be one sense class for each sense modality: for instance, a visual sense class containing just those quality classes which correspond to color spots. These sense classes are multidimensional arrays. The visual sense class, for instance, has five dimensions: two visual field coordinates, hue, brilliance, and saturation.

A METHOD OF IDENTIFYING SPURIOUS QUALITY CLASSES

Any genuine quality class will take its place in a sense class. There it will have many neighbors: genuine quality classes, genuinely similar to it. Some, at the edges of their sense classes or near gaps where their neighbors have failed to be constructed, will have fewer neighbors than the rest. Still, almost all genuine quality classes will be genuinely similar to many other quality classes. Genuine quality classes may also be spuriously similar to other quality classes, spurious or genuine.

A spurious quality class, on the other hand, cannot be included in any but spurious similarity circles. So it can be spuriously, but never genuinely, similar to other quality classes, spurious or genuine. But there is no evident reason why it should be more susceptible to spurious similarity than a genuine quality class. I conclude that it is likely to be similar to relatively few other quality classes, or to none.

Let the *neighborliness* of a quality class be the number of quality classes to which it is similar. Let a *hermit* be any quality class with low neighborliness; more precisely, any quality class such that no more than a specified small fraction of all other quality classes are less neighborly than it is. This fraction is the fraction of hermits among all the quality classes. If it were set at .1, say, the hermits would be defined as the least neighborly tenth of the quality classes.

I have argued that if the fraction is set properly, then we can expect almost all and only hermits to be spurious. So if we purge the hermits from the population of quality classes we will get rid of few genuine quality classes and most spurious ones. That is my proposal.

VARIANTS OF THE METHOD

We might want to guard strongly against purging genuine quality classes. We might set a low fraction of hermits, or we might even decide to remove only *total hermits:* quality classes with neighborliness 1, similar only to themselves. This is the most cautious version of my proposal. If spurious similarity is rare, plenty of spurious quality classes might be total hermits. But few genuine quality classes – even corner members of low-dimensional sense classes – would ever lose all their neighbors.

Next, notice that the fewer spurious similarities there are present, the more sharply do genuine and spurious quality classes contrast in neighborliness, so the more improvement we can get by removing hermits. It would be helpful to identify and remove spurious similarities before we identify and remove hermits. We can do so as follows. Let the *bypass distance* of a similarity between quality classes A and B be the number of steps in the shortest chain of similarities from A to B which would remain after the similarity between A and B had been removed. (If no chain would remain – that is, if the similarity connects what otherwise would have been two sense classes – let the bypass distance be infinite.) Let a *shortcut* be any similarity with high bypass distance; more precisely, any similarity such that no more than a specified small fraction of all other similarities have higher bypass distance than it has. (Let a *total shortcut* be any similarity with infinite bypass distance.) Since spurious similarities occur at random, they will often connect quality classes which would otherwise be far apart, perhaps even in different sense classes. So if the fraction of shortcuts is set properly, then we can expect almost all and only shortcuts to be spurious.

Therefore we might police the construction in two stages: first remove those similarities which are shortcuts (or total shortcuts, if we want to be cautious), then remove those quality classes which are hermits. Or we might go deeper. Instead of removing the shortcuts and hermits themselves, we might identify and remove those similarity circles which turned out to yield shortcuts or hermits, and afterwards use the surviving similarity circles to repeat the construction of quality classes and similarities.

Finally, I have been supposing so far that the fractions of shortcuts and hermits have somehow been chosen outside the construction. But we might let them be determined within the construction, making them depend on the statistics of bypass distance and neighborliness. We might take the highest bypass distance for which frequency in the frequency distribution of bypass distances reaches a local maximum, and let the shortcuts be those similarities with bypass distance falling more than a certain amount above that. Likewise we might take the lowest neighborliness for which frequency in the frequency distribution of neighborliness reaches a local maximum, and let the hermits be those quality classes with neighborliness falling more than a certain amount below that.

15

Finitude and infinitude in the atomic calculus of individuals

WITH WILFRID HODGES

Nelson Goodman has asked[1] whether there is any sentence in the language of his calculus of individuals which says whether there are finitely or infinitely many atoms. More precisely: is there any sentence that is true (false) in every finite intended model, regardless of its size, but not in any infinite atomic intended model? We shall show that there is no such sentence. We cannot say that there are finitely (infinitely) many atoms unless 1) we say something more specific about the number of atoms, or 2) we enlarge the language by providing for infinite conjunctions or disjunctions, or 3) we enlarge the language by providing suitable new predicates.

The language of the calculus of individuals is first-order logic with the following vocabulary of predicates:

xoy	(x overlaps y)	Sxyz	(the sum of x and y is z)
$x < y$	(x is part of y)	Nxz	(the negate of x is z)
$x = y$	(x is identical with y)	Ax	(x is an atom)

The *definitional axioms* of the calculus of individuals are (the universal closures of):

First published in *Noûs* 2 (1968), 405–410. Reprinted with kind permission from Blackwell Publishers and from Wilfrid Hodges.

The authors thank David Kaplan for some helpful remarks.

1 In lectures on Goodman, *The Structure of Appearance* (Cambridge, Mass.: Harvard University Press, 1951).

D1. $x < y \equiv \forall z(zox \supset zoy)$
D2. $x = y \equiv \forall z(zox \equiv zoy)$
D3. $Sxyz \equiv \forall w(woz \equiv wox \lor woy)$
D4. $Nxz \equiv \forall w(w < z \equiv \sim wox)$
D5. $Ax \equiv \forall z(z < x \supset z = x)$

The *existential axioms* of the calculus of individuals are (the universal closures of):

E1. $\exists z(z < x \,\&\, z < y) \equiv xoy$
E2. $\exists z(Sxyz)$
E3. $\exists z(Nxz) \equiv \sim\forall w(wox)$

The *axiom of atomicity* is:

AA. $\forall x \exists y(Ay \,\&\, y < x)$

The *atomic calculus of individuals* (henceforth ACI) is the theory axiomatized by D1–D5, E1–E3, and AA. By examining Goodman's discussion of the calculus of individuals,[2] it is easy to verify that the definitional and existential axioms hold in every intended model. The axiom of atomicity holds in just those intended models – *atomic intended models* – in which everything consists entirely of atoms. We are not concerned with the remaining intended models, in which things may consist wholly or partly of infinitely divisible nonatomic stuff. All finite intended models are atomic.

For any positive number *n,* we can write a certain sentence saying that there are at least *n* atoms. (Later we will say exactly which such sentence it is to be.) Call these sentences *numerative sentences.* We shall prove the following *Normal Form Theorem:*

> *Any sentence in the language of ACI is equivalent in ACI to a truth-functional compound of numerative sentences, and there is an effective procedure for finding one such equivalent of any given sentence.*

2 *Structure of Appearance,* II, 4. We have adapted Goodman's treatment only by introducing the predicates S and N; Goodman uses the corresponding functors, defined by means of definite descriptions.

Given this theorem, our negative answer to Goodman's question is an easy corollary. Call a sentence *indiscriminate* if and only if it has the same truth value in every infinite atomic intended model and also in every finite intended model with sufficiently many atoms. We seek a sentence that is not indiscriminate, being true in every finite intended model but false in every infinite atomic intended model (or vice versa). But every numerative sentence is indiscriminate, being true in every finite or infinite intended model with more than some number of atoms. And since every negation of an indiscriminate sentence is indiscriminate, and every conjunction of indiscriminate sentences is indiscriminate, every truth-functional compound of numerative sentences is indiscriminate. Every sentence equivalent in ACI to an indiscriminate sentence is indiscriminate. Therefore every sentence is indiscriminate, and the sentence we seek does not exist.

The Normal Form Theorem has several other interesting consequences. For any positive number n, let ACI_n be the theory obtained from ACI by adding as further axioms the numerative sentence saying that there are at least n atoms and the negation of the numerative sentence saying that there are at least $n + 1$ atoms; and let ACI_∞ be the theory obtained from ACI by adding as further axioms every numerative sentence. Call these theories *numerative extensions* of ACI. Assuming that intended models come in all finite, and some infinite, sizes – *size* being number of atoms – each numerative extension of ACI is the theory of a nonempty class of intended models: ACI_n is the theory of all intended atomic models of size n, ACI_∞ is the theory of all infinite atomic intended models. (We can restate our negative answer to Goodman's question thus: ACI_∞ is not finitely axiomatizable.) In each numerative extension of ACI, every numerative sentence can be effectively proved or disproved. Therefore the Normal Form Theorem provides a decision procedure for each numerative extension of ACI. It follows that the numerative extensions of ACI are maximal consistent, and it is easy to show that they are the only maximal consistent extensions of ACI. It also follows that ACI is a semantically complete theory of the class of all atomic intended models: given any set of sentences consistent with ACI, it can be embedded in a maximal consistent extension of ACI which, by our previous result, is one of the numerative extensions of ACI and therefore is true in the atomic intended models of the appropriate size. Finally, provided we disregard those models in which identity receives a nonstandard interpretation,[3] it follows that any finite model of ACI, intended or not, is isomorphic to an intended model: it satisfies some ACI_n along with the intended models of size n, and any two finite models of the same maximal consistent theory with (standardly interpreted) identity are isomorphic. We have obtained these results about atomic intended models without ever saying what those are; it was sufficient to know that they are models of ACI, they are standard with respect to identity, and they come in all finite, and some infinite, sizes.

It only remains to prove the Normal Form Theorem. We obtain it as a special case of a stronger normal form theorem, applicable not only to sentences but to all formulas in the language of ACI.

If n is any positive number and X and Y are any disjoint finite sets of variables, let Z be the set of the alphabetically first n variables not in X or Y, and let $[n, X, Y]$ be the formula which is the existential closure with respect to all the variables in Z (in alphabetical order) of the conjunction (in alphabetical order) of all the following formulas:

1) every formula $A\alpha$ with α in Z;
2) every formula $\sim\alpha = \beta$ with α in Z and β in $Z - \{\alpha\}$;
3) every formula $\alpha o \beta$ with α in Z and β in X;
4) every formula $\sim\alpha o \beta$ with α in Z and β in Y.

Let $W = X \cup Y$, so that W is the set of variables occurring free in $[n, X, Y]$; then we call $[n, X, Y]$ a W-*numerative formula*. Now we can define a *numerative sentence* as a sentence $[n, \wedge, \wedge]$ for some n, where $\wedge$ is the empty set; that is, as a $\wedge$-numerative formula. Call a formula ϕ *normalizable* if and only if ϕ is effectively equivalent in ACI to a truth-functional compound of W-numerative formulas, where W is the set of variables free in ϕ. Our Normal Form Theorem, which

3 That would be taken for granted if we regarded = as a logical constant, as is customary. But, following Goodman (*Structure of Appearance*, II, 4, D2.044), we do not, but rather adopt as a nonlogical axiom D2, in which it is defined in terms of the nonlogical predicate **o**.

says that every sentence is normalizable, follows from the theorem:

Every formula is normalizable.

Proof: It is sufficient to prove that any formula whose only predicate is **o** is normalizable, since every formula is effectively equivalent in ACI to such a formula, by the definitional axioms of ACI. Let ϕ be any such formula with the set W of free variables. We prove that ϕ is normalizable by induction on the complexity of ϕ.

Case 1: ϕ is αoβ. Then ϕ is normalizable, being equivalent in ACI to $[1, \{\alpha, \beta\}, \wedge]$.

Case 2: ϕ is a truth-functional compound of normalizable formulas. Let Ψ^i be any one of these; by hypothesis, it is effectively equivalent in ACI to a truth-functional compound of V^i-numerative formulas with V^i a subset of W. But any V^i-numerative formula is effectively equivalent to a truth-functional compound of W-numerative formulas; add the variables in $W - V^i$ one at a time by repeated use of the logical equivalence of any numerative formula $[n, X, Y]$ to the disjunction of all the following formulas, where α is the alphabetically first variable not in $X \cup Y$:

1) $[n, X \cup \{\alpha\}, Y]$;
2) $[n, X, Y \cup \{\alpha\}]$;
3) every conjunction $[m, X \cup \{\alpha\}, Y]$ & $[n\text{-}m, X, Y \cup \{\alpha\}]$ with $0 < m < n$.

Case 3: ϕ is an existential or universal quantification of a normalizable formula. The case of a universal quantifier reduces to that of an existential quantifier in view of case 2; so assume that ϕ is $\exists\alpha\Psi$, with Ψ normalizable. It can be shown that Ψ is effectively equivalent to a disjunction $\chi^1 \vee \ldots \vee \chi^k$, in which each formula χ^i is a conjunction of formulas χ_v^i indexed by all subsets of V of W and each formula χ_v^i is the conjunction of a formula of form 1, 2, or 3 and a formula of form 4, 5, or 6.

1) $\sim[m + 1, V \cup \{\alpha\}, W - V]$ with $m = 0$
2) $[m, V \cup \{\alpha\}, W - V]$ & $\sim[m + 1, V \cup \{\alpha\}, W - V]$
3) $[m, V \cup \{\alpha\}, W - V]$
4) $\sim[n + 1, V, (W - V) \cup \{\alpha\}]$ with $n = 0$
5) $[n, V, (W - V) \cup \{\alpha\}]$ & $\sim[n + 1, V, (W - V) \cup \{\alpha\}]$
6) $[n, V, (W - V) \cup \{\alpha\}]$

Intuitively, each assignment of values to the variables in W partitions the atoms into cells C_v indexed by all subsets V of W: C_v contains just those atoms which overlap the value of every variable in V but not the value of any variable in $W - V$. χ_v^i says that C_v contains exactly (or at least) m atoms of α and exactly (or at least) n other atoms, with $m, n \geq 0$. It is sufficient to show that each $\exists \alpha \chi^i$ is normalizable, since ϕ is plainly equivalent to the disjunction of these. Consider two subcases.

Case 3a: For every subset V of W, the first conjunct of χ_v^i has the form 1. Then $\exists \alpha \chi^i$ is inconsistent with ACI, and hence is equivalent in ACI to the conjunction of every formula $\sim[1, V, W - V]$ with V a subset of W.

Case 3b: Otherwise. Then $\exists \alpha \chi^i$ is equivalent in ACI to the conjunction of the formulas $\exists \alpha \chi_v^i$. It is sufficient to show that each of these is normalizable; consider any one. If the conjuncts of χ_v^i have the forms 1 and 4, $\exists \alpha \chi_v^i$ is equivalent in ACI to $\sim[1, V, W - V]$. If the conjuncts of χ_v^i have the forms 1 and 5, 2 and 4, or 2 and 5, $\exists \alpha \chi_v^i$ is equivalent in ACI to $[m + n, V, W - V]$ & $\sim[m + n + 1, V, W - V]$. Otherwise $\exists \alpha \chi_v^i$ is equivalent in ACI to $[m + n, V, W - V]$.

This completes the proof.

16

Nominalistic set theory

INTRODUCTION

By means that meet the standards of nominalism set by Nelson Goodman (in [1], section II, 3; and in [2]) we can define relations that
behave in many ways like the membership relation of set theory.
Though the agreement is imperfect, these pseudo-membership relations seem much closer to membership than to its usual nominalistic
counterpart, the part-whole relation. Someone impressed by the
diversity of set theories might regard the theories of these relations as
peculiar set theories; someone more impressed by the non-diversity
of the more successful set theories – ZF and its relatives – might prefer not to. This verbal dispute does not matter; what matters is that
the gap between nominalistic and set-theoretic methods of construction is narrower than it seems.

PRELIMINARIES

A finitistic nominalist's world might consist of an enormous hypercu-
bical array of space-time points, together with all wholes composed of
one or more of those points. Each point in the array is next to certain
others; nextness is a symmetric, irreflexive, intransitive relation among
the points. We can (but the nominalist cannot) describe the array of

First published in *Noûs* 4 (1970), 225–240. Reprinted with kind permission from
Blackwell Publishers.

points and the nextness relation more precisely by stipulating that the points can be placed in one-to-one correspondence with all the quadruples $\langle x, y, z, t \rangle$ of non-negative integers less than or equal to some very large integers x_{max}, y_{max}, z_{max}, t_{max}, respectively in such a way that one point is next to another iff the corresponding quadruples are alike in three coordinates and differ by exactly one in the remaining coordinate. We take this array for the sake of definiteness; but what follows does not depend on the exact shape and structure of the array. Large orderly arrays of various other sorts would do as well.

We will employ two primitive predicates:

$$x \text{ } is \text{ } part \text{ } of \text{ } y$$
$$x \text{ } is \text{ } next \text{ } to \text{ } y.$$

Both appear in [1] (sections II, 4 and X, 12), though not as primitive; so we can take them to be safely nominalistic. We will be concerned to define other predicates by means of these two primitives and first-order predicate logic with identity. (We will use English, but in such a way that it will be clear how to translate our definitions into logical notation.)

Using the part-whole predicate, we can define the various predicates of the calculus of individuals. We shall use them freely henceforth. In particular, let us define an *atom* as something having no parts except itself, so that atoms are the points in our array; and let us call x an *atom of* y iff x is an atom and x is part of y. Let us also say that x is the *universe* iff every atom is part of x.

From time to time we shall use set theory, but only in an auxiliary role. Set-theoretical definitions will not be part of our nominalistic construction. When we speak simply of sets or membership, we are to understand that standard sets or membership (say, in the sense of ZF) are meant. When we wish rather to speak of our nominalistic sets or membership, we will use subscripts or the prefix "pseudo-".

Let us say that x *touches* y iff some atom of x is next to some atom of y.

Let us call x the *interior* of y iff the atoms of x are all and only those atoms of y that are not next to any atoms except atoms of y. Not everything has an interior; a single atom, or a one-dimensional string of atoms, or a two-dimensional sheet of atoms, or a three-dimensional solid of atoms does not. Only the universe is identical with its interior.

Let us call x the *closure* of y iff the atoms of x are all and only those atoms that either are or are next to atoms of y. Everything has a closure. Only the universe is identical with its closure.

We might guess that everything that has an interior is the closure of its interior and that everything is the interior of its closure. Both guesses are wrong. Let us call something *stable* iff it has an interior and is the closure of its interior, and *solid* iff it is the interior of its closure. Not everything that has an interior is stable: if x is stable, y has no interior, and x does not touch y, then x + y has an interior but is not stable. Not everything is solid: if x is solid and has an interior and y is an atom of the interior of x, then x−y is not solid; the interior of the closure of x−y is x, since y is restored in taking the closure and remains when we take the interior. Note that the interior of any stable thing is solid and the closure of any solid thing is stable.

Let us call x and y *almost identical* iff each of the differences x-y and y-x either does not exist or has no interior. The relation of almost-identity is reflexive and symmetric but not transitive.

Let us call x *connected* iff, whenever y is a proper part of x, y touches x-y. The universe is connected. Let us call x *well-connected* iff the interior of x is connected. Let us call x a *maximal connected part of* y iff x is a connected part of y but x is not a part of any other connected part of y.

If x is the interior of a connected thing y, let us call y a *connection* of x and let us call x *connectible*. All solid connected things are connectible, since the closure of such a thing is one connection of it. With a few exceptions, solid disconnected things are connectible. We can make a connection of x by taking its closure and adding strings of atoms running from one maximal connected part of the closure of x to another. This can be done provided there are places to put the connecting strings such that every atom of the strings and every non-interior atom of the closure of x remains a non-interior atom of the closure-plus-strings. Then when we take the interior of the closure-plus-strings, the strings disappear and we get the interior of the closure – that is, since x is solid, we get x.

Something x could be solid but not connectible if, for instance, x was flattened out along some of the edges of our array. It could happen then that every non-interior atom of the closure of x was next to only one atom other than the atoms of the closure. Then we could not attach strings without converting an atom at the point of attachment from a non-interior atom of the closure to an interior atom of the closure-plus-strings. Since any such atom will be left behind when we take the interior, we will not recover x; so the closure-plus-strings is not a connection of x. Such cases, however, seem to occur only under rather special conditions.

MEMBERSHIP$_1$

Let us call x a *member$_1$* of y iff x is a maximal connected part of y. The theory of membership$_1$ resembles the usual theory of membership: that is, set theory. We have a principle of extensionality:

E1. If all and only members$_1$ of x are members$_1$ of y, then x and y are identical.

We have a restricted principle of comprehension:

C1. There is something z having as its members$_1$ all and only those things x such that __x__, provided that: (1) whenever __x__, x is connected; (2) no two objects x such that __x__ touch; (3) there is something x such that __x__.

(Here and henceforth, "__x__" is schematic for any formula of our language in which "x" appears free.) We also have a restricted principle of *Aussonderung:*

A1. There is something z having as its members$_1$ all and only those members$_1$ x of y such that __x__, provided that there is at least one such member$_1$ of y.

Indeed, for any non-empty set S of connected, non-touching things, there is a unique thing z whose members$_1$ are all and only the members of S: namely, the sum of the members of S.

Like membership, and unlike the part-whole relation, membership$_1$ is non-reflexive. Disconnected things are not members$_1$ of themselves or of anything else. Only connected things are members$_1$ of anything, and they are their own sole members$_1$ (*cf.* Quine's identification of individuals with their unit sets in [3]). Like the part-whole relation, and unlike membership, membership$_1$ is transitive. But in a trivial way: if x is a member$_1$ of y and y is a member$_1$ of z, then x is a member$_1$ of z because x and y are identical.

Like the part-whole relation, and unlike membership, membership$_1$ satisfies Goodman's criterion for a nominalistic generating relation: no distinction of entities without distinction of content ([2], section 2). It is not clear to me whether membership$_1$ is what Goodman calls a "generating relation"; but if it is one, then it is a nominalistic one. Let the *content$_1$* of x be the set of things which bear the ancestral of membership$_1$ to x but have no members$_1$ different from themselves. (This set-theoretical definition of a set is, of course, no part of our nominalistic construction.) No two different things have the same content$_1$.

MEMBERSHIP$_2$

The members$_1$ of something cannot touch, since if two things touch, they cannot both be maximal connected parts of anything. However, let us call x a *member$_2$* of y iff x is the closure of a maximal connected part of y and every maximal connected part of y is solid. Then for any non-empty set S of stable, well-connected, non-overlapping things, even if these things touch, there is a unique thing whose members$_2$ are all and only the members of S: namely, the sum of the interiors of members of S.

We have a principle of restricted extensionality, applying only to those things that have members$_2$:

E2. If all and only members$_2$ of x are members$_2$ of y and if x and y have members$_2$, then x and y are identical.

We have principles of restricted comprehension and restricted *Aussonderung:*

190

C2. There is something z having as its members$_2$ all and only those things x such that __x__, provided that: (1) whenever __x__, x is stable and well-connected; (2) no two objects x such that __x__ overlap; (3) there is something x such that __x__.

A2. There is something z having as its members$_2$ all and only those members$_2$ x of y such that __x__, provided that there is at least one such member$_2$ of y.

Any stable connected thing is a member$_2$ of something, but nothing else is. Anything whose maximal connected parts are solid has at least one member$_2$, but nothing else does. The proviso that a pseudo-set must consist of solid maximal connected parts is required to ensure extensionality.[1] Let x consist of solid maximal connected parts; let y be an atom of the interior of x; then exactly the same things are closures of maximal connected parts of x and of x-y. But x-y has a non-solid maximal connected part and hence has no members$_2$. Note that a pseudo-set consisting of solid maximal connected parts may itself not be solid.

Membership$_2$ is almost, but not quite, irreflexive. Only one thing is a member$_2$ of itself: the universe. Membership$_2$ is almost intransitive, with the same exception: if x is a member$_2$ of y and y is a member$_2$ of z, then x is not a member$_2$ of z unless x and y both are the universe.

Though nominalistically defined, membership$_2$ is not a nominalistic generating relation according to Goodman's criterion. Let the *content*$_2$ of x be the set of things which bear the ancestral of membership$_2$ to x but have no members$_2$ different from themselves. Then there are distinct things having the same content$_2$. For instance, if x and y are any two solid connected things such that the closures, the closures of the closures, etc. of x and y are solid, then x and y have exactly the same content$_2$: the unit set of the universe.

Not only the letter but the spirit of Goodman's criterion is violated. We get different entities corresponding to different ways of dividing up the same stuff, provided we consider only divisions of it into stable, well-connected, non-overlapping parts. Let the result of

1 I owe this observation to Oswaldo Chateaubriand.

enclosing a list in braces denote the thing having as members$_2$ all and only the listed things. Suppose we have three things x, y, and z which touch each other but do not overlap. Assume x, y, z, x + y, y + z, x + z, x + y + z all are stable and well-connected; and assume x, y, and z do not together exhaust the universe. Then we have the following six distinct things:

$$x + y + z \quad \{x + y + z\} \quad \{x, y, z\}$$
$$\{x + y, z\} \quad \{x, y + z\} \quad \{x + z, y\}$$

Yet we have not matched the standard set theorist's power to generate entities. We do not have, for instance, these:

$$\{x + y, y + z\} \quad \{x, x + y, x + y + z\}$$

If land were made out of our atoms, we could now handle one of Goodman's problematic sentences in [1] (section II, 3): "At least one group of lots into which it is proposed to divide this land violates city regulations."

We can thus have touching pseudo-members, but not overlapping ones. We might proceed further: call x a *member*$_2^n$ of y iff x is a member$_2$ of y; call x a *member*$_2^{n+1}$ of y iff x is a member$_2$ of a member$_2^n$ of y. (This is a definition-schema for a sequence of 2-place predicates, not a recursive specification for one 3-place predicate.) In this way we could get things with overlapping pseudo-members, provided they did not overlap too much. The higher we choose n, the more overlap we can tolerate but the stronger restrictions of other sorts we must build into the comprehension schema. Moreover, we will never be able to get something such that one of its pseudo-members is part of another one.

MEMBERSHIP$_3$

Let us put aside the task of providing for touching or overlapping pseudo-members and turn to the opposite task of providing for disconnected pseudo-members. Our plan will be to provide strings to tie together the disconnected parts of each pseudo-member.

Let us call x a *member*$_3$ of y iff x is the interior of a maximal connected part of y. For any non-empty set S of connected, solid things whose closures do not touch, there is a unique thing z whose members$_3$ are all and only the members of S: namely, the sum of the closures of the members of S. Also there is something z with no members$_3$: namely, anything with no interior. This memberless$_3$ thing is not unique, but it almost is; any two memberless$_3$ things are almost identical. For any connectible but disconnected thing x, there is something z whose sole member$_3$ is x: namely, any connection of x. It is not unique, but any two connections of x are almost identical. Finally, for any set S of connectible things whose closures do not touch (except in certain exceptional cases to be discussed), there is something z whose members$_3$ are all and only the members of S: namely, a sum of connections of the members of S chosen in such a way that no two of the connections touch. This thing z is not in general unique, but any two such things are almost identical.[2]

Trouble arises when the members of S are so crowded that there is no room for all the strings required to connect the disconnected members of S. In other words, there may be no way to choose connections of members of S so that no two of the connections touch. If so, there will be nothing whose members$_3$ are all and only the members of S.

In the worst case, suppose x and y are members of S, x is hollow, y is disconnected, part of y is inside x, and part of y is outside x. Then no matter how we connect y, the connection of y will overlap – and hence touch – x. Or suppose x is bottle-shaped, and suppose many disconnected members of S each have a part inside x and a part outside. They will all have to be connected by strings through the neck of x and there may not be room for that many strings to pass through the neck without touching each other or touching the closure of x.

The problem of crowding will not arise if S meets the following condition: there is something y such that some connection of any

2 We might overcome this weakening of extensionality if we had a predicate *precedes* such that for anything z there was a unique thing w having exactly the same members$_3$ as z and preceding everything else having exactly those members$_3$. Then we could take w as *the* pseudo-set with those pseudo-members, taking x to be a *member$_{3+}$* of y iff x is a member$_3$ of y and y precedes everything else having exactly those members$_3$. I do not know how to define such a precedence-relation, however, without taking some further primitives.

member of S is part of y and such that no connected part of y overlaps more than one member of S. (This condition also implies that the closures of members of S do not touch.) Unfortunately, it is complicated and not as distant as might be desired from the mere statement that there is something whose members$_3$ are all and only the members of S. Perhaps some simpler sufficient condition for non-crowding can be found.

We have a weakened principle of extensionality, a restricted principle of comprehension, and an unrestricted principle of *Aussonderung*:

E3. If all and only members$_3$ of x are members$_3$ of y, then x and y are almost identical.

C3. There is something z having as members$_3$ all and only those things x such that __x__, provided that: (1) whenever __x__, x is connectible; and (2) there is something y such that some connection of anything x such that __x__ is part of y and such that no connected part of y overlaps more than one thing x such that __x__.

A3. There is something z having as members$_3$ all and only those members$_3$ x of y such that __x__.

Membership$_3$ is almost irreflexive and intransitive. The exception is again the universe, which is its own sole member$_3$. Nothing else is a member$_3$ of itself. If x is a member$_3$ of y and y is a member$_3$ of z, then x is not a member$_3$ of z unless x, y, and z are all the universe. Membership$_3$ is not a nominalistic generating relation according to Goodman's condition. Let the *content$_3$* of x be the set of things which bear the ancestral of membership$_3$ to x but have no members$_3$ different from themselves; then different things can have the same content$_3$. The content$_3$ of the universe is the unit set of the universe; the content$_3$ of anything else turns out to be a set of one or more of the almost identical memberless$_3$ (because interiorless) things.[3]

Among the disconnected things which may be members$_3$ are pseudo-sets with more than one member$_3$. With membership$_3$ we begin

3 If we started with an infinite array of points, some things would have empty content$_3$.

194

to get something like the ordinary system of cumulative types. Suppose x and y are solid and connected; and suppose they are rather small and far apart so that there is no problem of crowding. Let the result of enclosing a list in braces now denote an arbitrarily chosen one of the things having as members$_3$ all and only the listed things. (In view of the arbitrary choice, this notation cannot be part of our construction.) Then we have, for instance, the following distinct things:

x	$\{x\}$	$\{\{x\}\}$	$\{\{\{x\}\}\}$
$\{x, y\}$	$\{\{x, y\}\}$	$\{\{x\}, y\}$	$\{\{x\}, \{y\}\}$
$x + y$	$\{x\} + y$	$\{x\} + \{y\}$	$\{x + y\}$

However, we do not have these:

$$\{x, \{x\}\} \qquad \{x, \{x, y\}\} \qquad \{x, x + y\}$$

Nor do we go on forever generating pseudo-sets of higher and higher type. We cannot: pseudo-sets are, after all, only sums of our finitely many atoms, so there can be only finitely many of them. Sooner or later crowding will set in and the proviso of C3 will fail.

If dogs and cats were made out of our atoms, did not touch one another, were connectible, and were uncrowded, then we could handle another of Goodman's problematic sentences: "There are more cats than dogs." ([1], section II, 3. Goodman handles this sentence by means of primitive predicates pertaining to size). We would say that there is something z such that (1) every member$_3$ of z has one cat and one dog as its sole members$_3$; (2) no cat or dog is a member$_3$ of two different members$_3$ of z; (3) every dog is a member$_3$ of a member$_3$ of z; but (4) not every cat is a member$_3$ of a member$_3$ of z. It would be possible to shorten the definition, but at the expense of obscuring the analogy to standard set theory.

In a similar vein we can do some arithmetic, correctly until crowding interferes. Call x and y *separate but equal* iff the closures of x and y do not touch and there is something z such that $x + y$ is part of z and such that every member$_3$ of z has one member$_3$ of x and one member$_3$ of y as its sole members$_3$. Call x and y *equal* iff something is separate but equal both to x and to y. Call x an *arithmetical sum* of y and z iff there is something w such that: (1) all members$_3$ of w are members$_3$ of x; (2) w is equal to y; and (3) x-w is equal to z. Call x an *arithmetical product* of y and z iff there are something w and something u such that: (1) all and only members$_3$ of members$_3$ of w are members$_3$ of u; (2) u is equal to x; (3) w is equal to y; and (4) every member$_3$ of w is equal to z. Call x a *zero* if x has no members$_3$. Call x a *one* if x has exactly one member$_3$. Call x a *two* iff x is an arithmetical sum of a one and a one; call x a *three* iff x is an arithmetical sum of a two and a one; and so on.

MEMBERSHIP$_4$

Membership$_2$ allows pseudo-members to touch, but requires them to be connected. Membership$_3$ allows pseudo-members to be disconnected but requires them not to touch. How might we combine the merits of the two? Let us call x a *member$_4$* of y iff x is the closure of the closure of the interior of a maximal connected part of y.

To see how this will work, suppose we want something z whose sole members$_4$ are x and y, disconnected things which touch but do not overlap each other. Only very well-behaved things are eligible to be members$_4$; we must assume that (1) x and y have interiors r and s respectively; (2) x and y are stable; (3) r and s have interiors t and u respectively; (4) r and s are stable; (5) t and u are connectible; and (6) t and u have connections v and w respectively such that v and w do not touch. (Assumption (6) is not unreasonable in view of the fact that the closures of t and u do not touch.) Now let z be v + w. When we take the maximal connected parts of z we get v and w. When we take the interiors of v and w we get t and u. When we take the closures of t and u we get r and s. Finally, when we take the closures of r and s – that is, the members$_4$ of z – we get x and y.

The theory of membership$_4$ resembles that of membership$_3$; we need not examine it at length. The principal difference is that the standards of eligibility for pseudo-membership are raised: if x is a member$_4$ of anything, x must be stable, the interior of x must be stable, and the interior of the interior of x must be connectible.

We could also allow disconnected pseudo-members to overlap slightly. Call x a *member$_4^0$* of y iff x is a member$_4$ of y; call x a *member$_4^{n+1}$* of y iff x is a closure of a member$_4^n$ of y. Of course we must pay for our increasing tolerance of overlap by elevating still further our other standards of eligibility for pseudo-membership; and even so we will not get pseudo-sets such that one pseudo-member is part of another.

MEMBERSHIP$_5$

Except for the almost identical zeros, everything has members$_3$; and except for the universe, everything has as its content$_3$ a set of one or more zeros. (The situation for membership$_4$ is analogous.) We have fallen into Pythagoreanism: everything except the universe is made out of pseudo-Zermelo numbers. This purity might please some. Others might wish to provide for *Urelements*: things without pseudo-members which are not all almost identical.

Call x an *Urelement* iff x is stable. Call x a *member$_5$* of y iff x is a member$_3$ of y and y is not an *Urelement*. We can easily destabilize something by adding to it another atom that does not touch the rest of it (unless it is almost identical with the universe, in which case no such atom is available). So for anything x (not almost identical with the universe) there is a non-*Urelement* z with the same members$_3$. Hence the members$_5$ of z are all and only the members$_3$ of x.

No *Urelement* or zero has members$_5$. Everything memberless$_5$ is either an *Urelement* or a zero. Zeros are not *Urelements* because they have no interiors. *Urelements* are not, in general, almost identical with one another. Some *Urelements*, however, are almost identical with other *Urelements*, with zeros, or with things having members$_5$.

The weakened principle of extensionality must of course be modified:

E5. If all and only members$_5$ of x are members$_5$ of y and if neither x nor y is an *Urelement*, then x and y are almost identical.

The anti-crowding proviso of the comprehension principle must be strengthened slightly. If connecting strings are required to form the pseudo-set of given things, they will do the destabilizing; but if not, we will need to make sure there is room for a destabilizing atom. It is sufficient to require that the thing y mentioned in the proviso of C3 not be almost identical with the universe.

The universe is no longer a pseudo-member of itself. Membership$_5$ is irreflexive and intransitive. The universe is anomalous in a new way: it is an *Urelement* too big to be a member$_5$ of anything. The same is true of any *Urelement* almost identical with the universe, and in general of any *Urelement* that is not connectible.

Defining the *content*$_5$ of x as usual, the content$_5$ of anything turns out to be a set of zeros and *Urelements*.

MEMBERSHIP$_6$

Membership$_6$ combines the merits of membership$_4$ and membership$_5$. Its definition, and that of membership$_6^n$ for any given n, are left to the reader.

MEMBERSHIP$_7$

We turn now to another pseudo-membership relation having two advantages over membership$_3$ and modifications thereof: (1) It handles disconnected pseudo-members without connecting strings, and hence without problems of crowding of the strings. We can even permit the pseudo-members of a pseudo-set to exhaust the entire universe, thus leaving no room for connecting strings. (2) It tolerates severe overlap of pseudo-members. We can even permit one pseudo-member of a pseudo-set to be a proper part of another. We buy these advantages, however, at a high price: we cannot have pseudo-sets of things that are too small, or that overlap only slightly, or that are themselves pseudo-sets.

Suppose we are given a set S of things. Members of S may be disconnected; they may together exhaust the universe; they may overlap; some may be proper parts of others. We may break the members of S into connected, non-overlapping fragments in such a way that every member of S is the sum of one or more of the fragments. Disconnected members of S will yield more than one fragment; overlapping members of S will yield fragments belonging to more than one member of S. We take interiors of the fragments, so that we may recover the fragments as closures of maximal connected parts of a pseudo-set; and we label the interiors of the fragments to carry information telling us which fragments to sum together to recover the members of S. All fragments which are part of any one member of S are to bear matching labels. Fragments which are part of two or more overlapping members of S must accordingly bear two or more labels.

In order to avoid crowding, we shall put each label of a fragment in a cavity hollowed out inside that fragment. Thus the label will be part of the interior of the fragment, made recognizable by removing some of the surrounding atoms. By making each label-holding cavity no larger than necessary, we can make sure that when we take the closure of the labeled interior of a fragment we will fill the cavities, and thereby erase the labeling and recover the entire fragment.

We may use as our labels connections of sums of closures of atoms, the atoms being sufficiently far apart so that the closures of closures of any two of them do not touch. Labels match when they are built around the same number of atoms; that is, when their interiors are equal.[4] Each label is surrounded by a skin and connected to the wall of its cavity by a string in such a way that the labeled interior of a fragment is connected, and each label is a maximal connected part of the interior of the labeled interior of its fragment.

Finally, we take as our pseudo-set corresponding to S the sum of labeled interiors of fragments. Under favorable conditions, such a thing will exist. We recover its pseudo-members according to the following definition: x is a *member₇* of y iff there is something z such that (1) z is a maximal connected part of the interior of some maximal connected part of y, and (2) x is the sum of the closures of all and only those maximal connected parts w of y such that the interior of some maximal connected part v of the interior of w is equal to the interior of z. (That is, z is a label and x consists of the closures of those maximal connected parts w of y such that w bears a label v that matches z.)

We do not yet have even our usual weakened principle of extensionality, since if x and y do not touch, y has an interior, and y bears no label, then x and x + y will have the same members₇ but will not be almost identical. Let us call x *minimal* iff no proper part of

---

4 Here we employ our previous definition of equality in terms of membership₃; and so our alternative to the direct use of connecting strings depends on a subsidiary use of them. However, it is unlikely that we will be troubled by crowding of strings. In comparing two interiors of labels there is no harm in letting a string run through regions occupied by members of S. We can ignore everything except the two things we are comparing and the other strings used in comparing them.

x has the same members$_7$ as x. Then we have weakened extensionality as follows:

E7. If all and only members$_7$ of x are members$_7$ of y and if x and y are minimal, then x and y are almost identical.

(Note that the only memberless$_7$ things that are minimal are the atoms, which are almost identical to each other; they are our empty pseudo-sets.) If we liked, we might revise the definition of pseudo-membership so that only minimal things have pseudo-members at all.

The principle of comprehension is restricted in a rather complicated way; we shall not attempt to state it. If there is to be a pseudo-set of the things x such that __x__, it must be possible to break those things into stable, well-connected fragments large enough to contain their labels. We will be unable to do this if some of the desired pseudo-members are too small, or if some of them overlap only slightly. The more pseudo-members there are, the larger must some of the labels be; the more overlapping pseudo-members share a common fragment, the more labels must be put inside that fragment. We are not limited by external crowding of strings, but instead by internal crowding of labels within the fragments.

We have an unrestricted principle of *Aussonderung*.

Defining the *content$_7$* of x as usual, we have different things with the same content$_7$; but only because of the weakening of extensionality, since the content$_7$ of anything is simply the set of its members$_7$. Nothing both has and is a member$_7$. We cannot construct pseudo-sets of higher than first type using membership$_7$; to do so it would be necessary to put a label inside a label, which is impossible.[5] So although membership$_7$ serves rather well for pseudo-sets of first type, because of its tolerance of overlap, it cannot be used in most set-theoretic constructions.

The set-theoretic constructions that cannot be carried out using membership$_7$, for lack of pseudo-sets of higher than first type, can

---

5 It might be possible to overcome this difficulty by using some other method of labeling, probably at the cost of worsening the problem of internal crowding.

often be replaced by constructions using both membership$_7$ and membership$_3$. A pseudo-set based on membership$_7$ gives us a representing relation, defined as follows: x *represents* y *with respect to* z iff y is a member$_7$ of z and for every maximal connected part w of z, w is part of y if and only if there is a maximal connected part v of the interior of w such that the interiors of v and x are equal. (That is: x matches the labels that unite the fragments of y in z.) We can use membership$_7$ in this way to obtain a representing relation for the things we are interested in, provided there is something z having those things among its members$_7$. Then instead of constructing pseudo-sets of high type out of the members$_7$ of z themselves, we can rather construct pseudo-sets of high type out of their representatives. We can do this using membership$_3$, which is good at dealing with high types. Membership$_3$ cannot handle overlap, but that does not matter. Since anything equal to a representative of a given thing also represents that thing (with respect to z), each member$_7$ of z has many representatives scattered around the universe. In constructing pseudo-sets of high type by means of membership$_3$ out of representatives of members$_7$ of z, we are free to use many representatives of the same thing, chosen in such a way as to avoid overlap.

Suppose we want to say that among the political divisions of California at all levels – counties, cities, precincts, school districts, assembly districts, etc. – more have Democratic majorities than have Republican majorities. (We may pretend that these political divisions are made out of our atoms.) We could not say this using our earlier methods because of overlap: a Democratic school district may overlap a Democratic assembly district and be a proper part of a Democratic county. But we may say this: there exist x, y, z, such that (1) every political division of California with a Democratic or Republican majority is a member$_7$ of z, (2) every political division of California with a Democratic or Republican majority is represented, with respect to z, by exactly one member$_3$ either of x or of y, (3) every member$_3$ of x represents, with respect to z, a political division of California with a Democratic majority, (4) every member$_3$ of y represents, with respect to z, a political division of California with a Republican majority, and (5) y is equal to something w having as members$_3$ some but not all members$_3$ of x and no other things.

We can extend our representing relation to sets of arbitrarily high cumulative type constructed out of members$_7$ of something z by means of the following recursive clause: x *represents* S *with respect to* z iff for each member y of S there is exactly one member$_3$ w of x, and for each member$_3$ w of x there is some member y of S, such that w represents y with respect to z. (This recursive specification of a relation involving sets, unlike our original definition of representation of individuals, is no part of our nominalistic construction.)

For example, if x and y are members$_7$ of z, the ordered pair of x and y is represented with respect to z by anything w having exactly two members$_3$ u and v such that (1) u has exactly one member$_3$ r, (2) v has exactly two members$_3$ s and t, (3) r and s both represent x with respect to z, and (4) t represents y with respect to z. Note that the two representatives r and s of x must be different.

For another example, x represents with respect to z the set of all sets of members$_7$ of z iff: (1) each member$_3$ of a member$_3$ of x represents a member$_7$ of z with respect to z, and (2) for anything y such that each member$_3$ of y represents a member$_7$ of z with respect to z, there is a member$_3$ w of x such that the members$_3$ of w and the members$_3$ of y represent, with respect to z, exactly the same members$_7$ of z.

In this way, until external crowding among representatives interferes, we can represent any set, no matter how high its type, whose content consists of members$_7$ of something z.

REFERENCES

[1] Goodman, Nelson, *The Structure of Appearance,* 2nd edition (Indianapolis: Bobbs-Merrill, 1966).

[2] ———, "A World of Individuals," reprinted in Benacerraf and Putnam, *Philosophy of Mathematics* (Englewood Cliffs, N.J.: Prentice-Hall, 1964).

[3] Quine, W. V., *Set Theory and Its Logic,* 2nd edition (Cambridge, Mass.: Harvard University Press, 1969).

17

Mathematics is megethology

Mereology is the theory of the relation of part to whole, and kindred notions. Megethology is the result of adding plural quantification, as advocated by George Boolos in [1] and [2], to the language of mereology. It is so-called because it turns out to have enough expressive power to let us express interesting hypotheses about the size of Reality. It also has the power, as John P. Burgess and A. P. Hazen have shown in [3], to simulate quantification over relations.

It is generally accepted that mathematics reduces to set theory. In my book *Parts of Classes,* [6], I argued that set theory in turn reduces, with the aid of mereology, to the theory of singleton functions. I also argued (somewhat reluctantly) for a 'structuralist' approach to the theory of singleton functions. We need not think we have somehow achieved a primitive grasp of some one special singleton function. Rather, we can take the theory of singleton functions, and hence set theory, and hence mathematics, to consist of generalisations about all singleton functions. We need only assume, to avoid vacuity, that there exists at least one singleton function.

But we need not assume even that. For it now turns out that if the size of Reality is right, there must exist a singleton function. All we need as a foundation for mathematics, apart from the framework of megethology, are some hypotheses about the size of Reality.

First published in this form in *Philosophia Mathematica* 3 (1993), 3–23. This paper consists in part of near-verbatim excerpts from David Lewis, *Parts of Classes* (Oxford, Blackwell, 1991). Reprinted with kind permission from *Philosophia Mathematica* and from Blackwell Publishers.

(Megethology can have no complete axiom system; and it would serve little purpose to fix upon some one official choice of an incomplete fragment. Rather we shall proceed informally, and help ourselves to principles of megethology as need arises. Only the less obvious principles will be noted.)

This article is an abridgement of parts of *Parts of Classes* not as it is, but as it would have been had I known sooner what I know now. It begins by repeating some material from the early sections of the book, mostly by means of near-verbatim excerpts. It mostly skips the philosophical middle sections. It gives a new presentation of some of the technical material from the late sections, simplified by using simulated quantification over relations. And it proves the new result that if the size of Reality is right, then there exists a singleton function.

One mereological notion is that of a *fusion* or *sum:* the whole composed of some given parts (see [5], II. 4). The fusion of all cats is that large, scattered chunk of cat-stuff which is composed of all the cats there are, and nothing else. It has all cats as parts. There are other things that have all cats as parts. But the cat-fusion is the least such thing: it is included as a part in any other one.

It does have other parts too: all cat-parts are parts of it, for instance cat-whiskers, cat-quarks. For parthood is transitive; whatever is part of a cat is thereby part of a part of the cat-fusion, and so must itself be part of the cat-fusion.

The cat-fusion has still other parts. We count it as a part of itself: an *improper* part, a part identical to the whole. But also it has plenty of *proper parts* – parts not identical to the whole – besides the cats and cat-parts already mentioned. Lesser fusions of cats, for instance the fusion of my two cats Magpie and Possum, are proper parts of the grand fusion of all cats. Fusions of cat-parts are parts of it too, for instance the fusion of Possum's paws plus Magpie's whiskers, or the fusion of all cat-tails wherever they be. Fusions of several cats plus several cat-parts are parts of it. And yet the cat-fusion is made of nothing but cats, in this sense: it has no part that is entirely distinct from each and every cat. Rather, every part of it overlaps some cat.

We could equivalently define the cat-fusion as the thing that overlaps all and only those things that overlap some cat. Since all and only overlappers of cats are overlappers of cat-parts, the fusion of all cats is the same as the fusion of all cat-parts. It is also the fusion of all cat-molecules, the fusion of all cat-particles, and the fusion of all things that are either cat-front-halves or cat-back-halves. And since all and only overlappers of cats are overlappers of cat-fusions, the fusion of all cats is the same as the fusion of all cat-fusions.

The *class* of all cats is something else. It has all and only cats as members. Cat-parts such as whiskers or cells or quarks are *parts of* members of it, but they are not themselves members of it, because they are not whole cats. Cat-parts are indeed members of the class of all cat-parts, but that's a different class. Fusions of several cats are *fusions of* members of the class of all cats, but again they are not themselves members of it. They are members of the class of cat-fusions, but again that's a different class.

The class of A's and the class of B's are identical only if the A's are all and only the B's; but the fusion of the A's and the fusion of the B's can be identical even when no A is a B. Therefore we learn not to identify the class of A's with the fusion of A's, and the class of B's with the fusion of B's, lest we identify two different classes with one single fusion.

A member of a member of something is not, in general, a member of it; but a part of a part of something is always a part of it. Therefore we learn not to identify membership with the relation of part to whole.

So far, so good. But I used to think, and so perhaps did you, that we learned more. We learned to distinguish two entirely different ways to make one thing out of many: the way that made one fusion out of many parts, *versus* the way that made one class out of many members. We learned that fusions and classes were two quite different kinds of things, so that no class was ever a fusion. We learned that the part-whole relation applies to individuals, not sets. We even learned to call mereology 'The Calculus of Individuals'!

All that was a big mistake. Just because a class isn't the mereological fusion of its members, we needn't conclude that it isn't a fusion. Just because one class isn't composed mereologically out of its many members, we needn't conclude that there must be some unmereological way to make one out of many. Just because a class doesn't have all and only its members as parts, we needn't conclude that it has no parts.

Mereology does apply to classes. They do have parts: their subclasses. (It remains to be seen whether they have other parts as well.) But for now we have this

First Thesis. *One class is part of another iff the first is a subclass of the second.*

To explain what the First Thesis means, I must hasten to tell you that my usage is a little bit idiosyncratic. By 'classes' I mean things that have members. By 'individuals' I mean things that are members, but do not themselves have members. Therefore there is no such class as the null class. I don't mind calling some memberless thing – some individual – the null *set*. But that doesn't make it a memberless class. Rather, that makes it a 'set' that is not a class. Standardly, all sets are classes and none are individuals. I am sorry to stray, but I must if I am to mark the line that matters: the line between the membered and the memberless. Besides, we had more than enough words. I can hijack 'class' and 'individual', and still leave other words unmolested to keep their standard meanings. As follows: a *proper class* is a class that is not a member of anything; a *set* is either the null set or else a class that is not a proper class.

My First Thesis, therefore, has nothing to say yet about the null set. It does not say whether the null set is part of any classes, nor whether any classes are part of the null set. I shall take up those questions later. Now that you understand what the First Thesis means, what can I say in its favour?

First, it conforms to common speech. It does come natural to say that a subclass is part of a class: the class of women is part of the class of human beings, the class of even numbers is part of the class of natural numbers, and so on. Likewise it comes natural to say that a hyperbola has two separate parts – and not to take that back when we go on to say that the hyperbola is a class of x-y pairs. The devious explanation of what we say is that we speak metaphorically, guided by an analogy of formal character between the part-whole relation and the subclass relation. The straightforward explanation is that subclasses just are parts of classes, we know it, we speak accordingly.

Second, the First Thesis faces no formal obstacles. We learned, rightly, that membership could not be (a special case of) the part-whole relation because of a difference in formal character. But the subclass relation and the part-whole relation behave alike. Just as a part of a part is itself a part, so a subclass of a subclass is itself a subclass; whereas a member of a member is not in general a member. Just as a whole divides exhaustively into parts in many different ways, so a class divides exhaustively into subclasses in many different ways; whereas a class divides exhaustively into members in only one way. We have at very least an analogy of formal character, wherefore we are free to claim that there is more than a mere analogy.

Finally, I hope to show you that the First Thesis will prove fruitful. Set theory is peculiar. It all seems so innocent at first! We need only accept that when there are many things, then also there is one thing – the class – which is just the many taken together. It's hard to object to that. But it turns out later that this many-into-one can't always work, on pain of contradiction – yet it's just as hard to object to it when it doesn't work as when it does. What's more, the innocent business of making many into one somehow transforms itself into a remarkable making of one into many. Given just one single individual, Magpie or Possum or the null set or what you will, suddenly we find ourselves committed to a vast hierarchy of classes built up from it. Not so innocent after all! This ontological extravagance is just what gives set theory its welcome mathematical power. But, like it or not, it's far from what we bargained for when we first agreed that many can be taken together as one. We could understand set theory much better if we could separate the innocent Dr. Jekyll from the extravagant and powerful Mr. Hyde. The First Thesis is our first, and principal, step toward that separation.

The First Thesis leaves it open that classes might have other parts as well, besides their subclasses. Maybe classes sometimes, or always, have individuals as additional parts: the null set, cat Magpie, Possum's tail (and with it all the tail-segments, cells, quarks, and what-not that are parts of Possum's tail). To settle the question, I advance this

Second Thesis. *No class has any part that is not a class.*

The conjunction of the First and Second Theses is our

Main Thesis. *The parts of a class are all and only its subclasses.*

But the Second Thesis seems to me far less evident than the First; it needs an argument. The needed premises are the First Thesis plus three more.

Division Thesis. *Reality divides into individuals and classes.**

Priority Thesis. *No class is part of any individual.*

Fusion Thesis. *Any fusion of individuals is itself an individual.*

Roughly speaking, the Division Thesis says that there is nothing else except individuals and classes. But that is not exactly right. If we thought that Reality divided exhaustively into animal, vegetable, and mineral, that would not mean that there was no such thing as a salt beef sandwich. The sandwich is no counterexample, because the sandwich itself divides: the beef is animal, the bread is vegetable, and the salt is mineral. Likewise, the Division Thesis permits there to be a mixed thing which is neither an individual nor a class, so long as it divides exhaustively into individuals and classes. I accept a principle of *Unrestricted Composition:* whenever there are some things, no matter how many or how unrelated or how disparate in character they may be, they have a mereological fusion. That means that if I accept individuals and I accept classes, I have to accept mereological fusions of individuals and classes. Like the mereological fusion of the front half of a trout plus the back half of a turkey, which is neither fish nor fowl, these things can be mostly ignored. They can be left out of the domains of all but our most unrestricted quantifying. They resist

---

* [Added 1996] The Division Thesis is badly worded. The meaning that I intended, and that is required by subsequent discussions here and in *Parts of Classes*, is better expressed as follows: everything is either an individual, or a class, or a fusion of an individual and a class. I thank Daniel Nolan for pointing out the problem to me.

concise classification: all we can say is that the salt beef sandwich is part animal, part vegetable, part mineral; the trout-turkey is part fish and part fowl; and the mereological fusion of Possum plus the class of all cat-whiskers is part individual and part class. Likewise, Reality itself – the mereological fusion of everything – is mixed. It is neither individual nor class, but it divides exhaustively into individuals and classes. Indeed, it divides into one part which is the most inclusive individual and another which is the most inclusive class.

(If we accept the mixed fusions of individuals and classes, must we also posit some previously ignored classes that have these mixed fusions as members? No; mixed fusions are forced upon us by the principle of Unrestricted Composition, but classes having mixed fusions as members are not forced upon us by any otherwise acceptable principle. Let us indulge our offhand reluctance to believe in them.)

All I can say to defend the Division Thesis, and it's weak, is that as yet we have no idea of any third sort of thing that is neither individual nor class nor mixture of the two. Remember what an individual is: not necessarily a commonplace individual like Magpie or Possum, or a quark, or a spacetime point, but anything whatever that has no members but is a member. If you believe in some remarkable non-classes – universals, tropes, abstract simple states of affairs, God, or what you will – it makes no difference. They're still individuals, however remarkable, so long as they're members of classes and not themselves classes. Rejecting the Division Thesis means positing some new and hitherto unheard-of disqualification from membership, applicable to things that neither are classes nor have classes as parts. I wouldn't object to such a novel proposal, if there were some good theoretical reason for it. But so long as we have no good reason to innovate, let conservatism rule.

The Priority Thesis and the Fusion Thesis reflect our vague notion that somehow the individuals are 'basic' and 'self-contained' and that the classes are somehow a 'superstructure'; 'first' we have individuals and the classes come 'later'. (In some sense. But it's not that God made the individuals on the first day and the classes not until the second.) Indeed, these two theses may be all the sense that we can extract from that notion. We don't know what classes are made of – that's what we want to figure out. But we do know what individuals are made of: various smaller individuals, and nothing else.

From the First Thesis, the Division Thesis, the Priority Thesis, and the Fusion Thesis, our Second Thesis follows.

> Proof. If the Thesis fails, some class x has a part y that is not a class. By the Division Thesis, y is either an individual or a mixed fusion, and either way, x has an individual as part. Let z be the fusion of all individuals that are part of x; then z is an individual, by the Fusion Thesis. Now consider the difference $x - z$ (the fusion of all parts of x that do not overlap z). Since $x - z$ has no individuals as parts, it is not an individual or a mixed fusion. By the Division Thesis, it must be a class. We now have that x is the fusion of class $x - z$ with an individual z. Since $x - z$ is part of x, and not the whole of x (else there wouldn't have been any z to remove), we have that the class $x - z$ is a proper part of the class x. So, by the First Thesis, $x - z$ must be a proper subclass of x. Then we have v, a member of x but not of $x - z$. According to standard set theory, we then have u, the class with v as its only member. By the First Thesis, u is part of x but not of $x - z$; by the Priority Thesis, u is not part of z; so u has some proper part w that does not overlap z. No individual is part of w; so by the Division Thesis, w is a class. By the First Thesis, w is a proper subclass of u. But u, being one-membered, has no proper subclass. This completes a *reductio*. QED

A consequence of our Second Thesis is that classes do not have the null set as part. Because it was a memberless member, we counted it as an individual, not a class; therefore it falls under our denial that individuals ever are parts of classes. (To be sure, it is *included* in any class, because all its members – all none of them – are members of that class. But it never can be a *sub*class if it is not even a class.) Were we hasty? Should we amend the Second Thesis, and the premises whence we derived it, to let the null set be a part of classes after all? I think not.

Or should we perhaps reject the null set? Is it a misguided posit, meant to streamline the formulation of set theory by behaving in peculiar ways? Again, I think not. Its behaviour is not, after all, so very peculiar. It is included in every class just because it lacks members – and lacking members is not so queer, all individuals do it. And it serves two useful purposes. It is a denotation of last resort for class abstracts that denote no nonempty class. And it is an individual of last resort: we can count on its existence in a way we could count on the existence of nothing else, and once we have it we can fearlessly build up the hierarchy of pure sets: the null set, the singleton of the null set, the singleton of the singleton of the null set, the set of the null set and its singleton, . . . *ad infinitum* and beyond, until we have enough modelling clay to build the whole of mathematics.

Or should we accept the null set as a most extraordinary individual, a little speck of sheer nothingness, a sort of black hole in the fabric of Reality itself? Not that either, I think.

We want a null set, but we needn't be ontologically serious about it. It's useful to have a name that's guaranteed to denote some individual, but it needn't be a special individual with a whiff of nothingness about it. An ordinary individual – *any* ordinary individual – will suffice. Any individual has the first qualification for the job – memberlessness. As for the second qualification, guaranteed existence, that is not really a qualification of the job-holder itself, rather it is a requirement on our method of choice. To guarantee that we'll choose some existing individual to be the null set, we needn't choose something that's guaranteed to exist. It's enough to make sure that we choose from among whatever things happen to exist. The choice is arbitrary. I make it – arbitrarily! – as follows: let the null set be the fusion of all individuals. If any individuals exist, this selects one of them for the job. We also get a handy name for one of the main subdivisions of Reality: if the null set is the fusion of all individuals, then by our Theses, the individuals are all and only the parts of the null set. It makes the null set omnipresent, and thereby respects our 'intuition' that it is no more one place than another. It's far from the notion that the null set is a speck of nothingness, and that's all to the good.

Our Main Thesis says that the parts of a class are all and only its subclasses. This applies, in particular, to one-membered classes: *unit classes,* or *singletons*. Possum's singleton has Possum as its sole member. It has no subclasses except itself. Therefore it is a mereological *atom:* it has no parts except itself, no proper parts. Likewise the singleton of Possum's singleton is an atom; and likewise for any other singleton.

211

Anything that can be a member of a class has a singleton: every individual has a singleton, and so does every set. The only things that lack singletons are the proper classes – classes that are not members of anything, and *a fortiori* not members of singletons – and those mixed things that are part individual and part class. And, of course, nothing has two singletons. So the singletons correspond one-one with the individuals and sets.

A class has its singleton subclasses as atomic parts, one for each of its members. Its larger parts (unless it is a singleton) are its non-singleton subclasses. A class is the union, and hence the fusion, of the singletons of its members. For example, the class of the two cats Possum and Magpie is the fusion of Possum's singleton and Magpie's singleton. The class whose three members are Possum, Magpie's singleton, and the aforementioned class is the fusion of Possum's singleton, Magpie's singleton's singleton, and the singleton of the aforementioned class. And so it goes.

Taking the notion of a singleton henceforth as primitive, and appealing to our several theses, we get these new definitions. Membership, hitherto primitive, shall be so no longer.

Classes are fusions of singletons.

Members of a class are things whose singletons are parts of that class.

(We must add that only classes have members. A mixed fusion has singletons as parts, but we probably would not want to say that it had members.)

Individuals are things that have no singletons as parts.

The fusion of all individuals, our choice to be the null set, is therefore the mereological difference, Reality minus all the singletons.

Sets are the null set, and all classes that have singletons.

Proper classes are classes that have no singletons.

The class of all sets that are non-self-members had better not be a set, on pain of Russell's paradox. Although it is indeed a non-self-member, that won't make it a self-member unless it is a set. So it isn't; it is a proper class, it has no singleton, and it cannot be a member of anything.

We dare not allow a *set* of all sets that are non-self-members, but there are two alternative ways to avoid it. One way is to restrict composition: we have all the sets that are non-self-members and we have a singleton of each of these sets, but somehow we have no fusion of all these singletons, so we have no class of all sets that are non-self-members. But there is no good independent reason to restrict composition. Mereology *per se* is unproblematic, and not to blame for the set-theoretical paradoxes; so it would be unduly drastic to stop the paradoxes by mutilating mereology, if there is any other remedy. (Just as it would be unduly drastic to solve problems in quantum physics by mutilating logic, or problems in the philosophy of mind and language by mutilating mathematics.) The better remedy, which I have adopted, is to restrict not composition, but rather the making of singletons. We *can* fuse all the singletons of all the sets that are non-self-members, thereby obtaining a class, but this class does not in turn have a singleton; it is proper. Unlike composition, the making of singletons is ill-understood to begin with, so we should not be surprised or disturbed to find that it needs restricting.

I do not say, note well, that we must posit the proper classes for the sake of their theoretical utility. We have them willy-nilly, be they useful or be they useless. We do not go out of our way to posit them. We just can't keep them away, given our Main Thesis and Unrestricted Composition.

The proper classes aren't much use, in fact. For George Boolos has argued convincingly in [1] and [2] that we do not require proper classes in order to formulate powerful systems of set theory. We can get the needed power instead by resorting to irreducibly plural quantification, something well-known to us as speakers of ordinary language. Suppose we say 'If there are some people such that each of Peter's parents is one of them, and every parent of one of them is one of them, then each of Peter's ancestors is one of them'; here it seems, *prima facie*, that we are quantifying just over people – not over classes. (And not over any class-like entities that differ from classes only in name.) And if we say 'Some things are all and only the classes that

213

are non-self-members', this seems to be a trivial truth (given that there are such classes), not a paradoxical assertion of the existence of the Russell class. (Note that this time I said 'classes', not 'sets', to block the solution which invokes a proper class.) I join Boolos in concluding that what's true is just what seems to be true. Plural quantification is not class quantification in sheep's clothing. It is innocent of set theory, except of course when we quantify plurally over classes themselves. I shall make free with plural quantification in presenting mereology. In fact, I've done so already. When I said 'whenever there are some things they have a fusion', my plural quantifier could not have been read without loss as a class quantifier or as a substitutional 'quantifier'. I meant that *whenever* there are some things, regardless of whether they are members of some class, and regardless of whether they are the satisfiers of some formula, they have a fusion.

So we could rebuild set theory within mereology, if only we had the primitive notion of singleton. But that, I fear, is a tall order.

Cantor taught that a set is a 'many, which can be thought of as one, *i.e.,* a totality of definite elements that can be combined into a whole by a law'. To this day, when a student is first introduced to set theory, he is apt to be told something similar. He is told that a set is formed by combining or collecting or gathering several objects, or by thinking of them together. Maybe also he will be given some familiar examples: Halmos's textbook mentions packs of wolves, bunches of grapes, and flocks of pigeons. But after a time, the unfortunate student is told that some classes – the singletons – have only a single member. Here is just cause for student protest, if ever there was one. This time, he has no 'many'. He has no elements or objects to be 'combined' or 'gathered together' into one, or to be 'thought of together as one'. Rather, he has just one single thing, the element, and he has another single thing, the singleton, and nothing he was told gives him the slightest guidance about what the one thing has to do with the other. Nor did any of those familiar examples concern single-membered sets. His introductory lesson just does not apply.

He might think: whatever it is that you do, in action or in thought, to make several things into a class, just do that same thing to a single thing and you make it into a singleton. How do you make several paintings into an art collection? Maybe you make a plan, you buy the paintings, you hang them in a special room, you even publish a catalogue. If you do the same thing, but your money runs out after you buy the first painting on your list, you have a collection that consists of a single painting. – But this thought is worse than useless. For all those allusions to human activity in the forming of sets are a bum steer. Sooner or later our student will hear that there are countless classes, most of them infinite and miscellaneous, so that the vast majority of them must have somehow got 'formed' with absolutely no attention or assistance from us. Maybe we've formed a general concept of classes, or a theory of them, or some sort of sketchy mental map of the whole of set-theoretical Reality. Maybe we've formed a few mental representations of a few very special classes. But there just cannot be anything that we've done to all the classes one at a time. The job is far too big for us. Must set theory rest on theology? – Cantor thought so!

We were told nothing about the nature of the singletons, and nothing about the nature of their relation to their elements. That might not be quite so bad if the singletons were a very special case. At least we'd know about the rest of the classes. But since all classes are fusions of singletons, and nothing over and above the singletons they're made of, our utter ignorance about the nature of the singletons amounts to utter ignorance about the nature of classes generally. We understand how bigger classes are composed of their singleton atoms. That's the easy part: just mereology. *That's* where we get the many into one, the combining or collecting or gathering. Those introductory remarks (apart from the misguided allusions to human activity) introduced us only to the *mereology* in set theory. But as to what is distinctively set-theoretical – the singletons that are the building blocks of all classes – they were entirely silent. Dr. Jekyll was there to welcome us. Mr. Hyde kept hidden. What do we know about singletons when we know only that they are atoms, and wholly distinct from the familiar individuals? What do we know about other classes, when we know only that they are composed of these atoms about which we know next to nothing?

Set theory has its unofficial axioms, traditional remarks about the nature of classes. They are never argued, but are passed along heedlessly from one author to another. Some are acceptable enough, I suppose: they do nothing to characterise the classes positively, but limit themselves to a *via negativa*. They say that such things as cats and quarks and spacetime regions should come out as individuals, not classes or mixed fusions.

Other unofficial axioms are bolder. One of them says that the classes are nowhere: they are outside of space and time. But why think this? Because we never see them or stumble over them? But maybe they are invisible and intangible. Maybe they can share their locations with other things. Maybe Possum's singleton is just where Possum is; maybe every singleton is just where its member is. Since members of singletons occupy extended spatiotemporal regions, and singletons are atoms, that would have to mean that something can occupy an extended region otherwise than by having different parts that occupy different parts of the region, and that would certainly be peculiar. But not more peculiar, I think, than being nowhere at all – we get a choice of equal evils, and cannot reject either hypothesis by pointing to the repugnancy of the other. I don't say the classes are in space and time. I don't say they aren't. I say we're in the sad fix that we haven't a clue whether they are or whether they aren't. We go much too fast from not knowing whether they are to thinking we know they are not.

Another unofficial axiom says that classes have nothing much by way of intrinsic character. That's not quite right: to be an atom, or to be a fusion of atoms, or to be a fusion of seventeen atoms, are matters of intrinsic character. However, these are not matters that distinguish one singleton from another, or one seventeen-membered class (a seventeen-fold fusion of singletons) from another. Are all singletons exact intrinsic duplicates? Or do they sometimes, or do they always, differ in their intrinsic character? If they do, do those differences in any way reflect differences between the character of their members? Do they involve any of the same qualities that distinguish individuals from one another? Again we haven't a clue.

Sometimes our offhand opinions about the nature of classes don't even agree with one another. When Nelson Goodman ([5], II.2) finds the notion of classes 'essentially incomprehensible' and refuses to 'use apparatus that peoples his world with a host of ethereal, platonic, pseudo entities' we should ask which are they: ethereal or platonic? The ether is everywhere, and one bit of it is pretty much like another; whereas the forms are nowhere, and each of them is unique. Ethereal entities are 'light, airy or tenuous', says the dictionary, whereas the forms are changeless and most fully real (whatever that means). If we knew better whether the classes were more fittingly called 'ethereal' or 'platonic', that would be no small advance!

I suppose we could bear up under our utter ignorance of the character and whereabouts (or lack thereof) of the singletons. Who ever said we could know everything? But there is worse to come. Because we know so little about the singletons, we are ill-placed even to begin to understand the relation of a thing to its singleton. We know what to call it, of course − membership − but that is all. Is it an external relation, like a relation of distance? An internal relation, like a relation of intrinsic similarity or difference? A combination of the two? Something else altogether?

It's no good saying that a singleton has x as its member because it shares the location of x. If singletons do share the location of their members, then x and x's singleton and x's singleton's singleton all three share a location; so x shares that location as much with one singleton as with the other. It's no good saying that a singleton has x as its member because of some sort of similarity between the singleton and x. For two perfect duplicates may have different singletons. It's no good saying that a singleton has x as its member because it has the property: being the singleton of x. That's just to go in a circle. We've named a property; but all we know about the property that bears this name is that it's the property, we know not what, that distinguishes the singleton of x from all other singletons.

It seems that we have no alternative but to suppose that the relation of member to singleton holds in virtue of qualities or external relations of which we have no conception whatsoever. Do we really understand what it means for a singleton to have a member?

Singletons, hence all classes, and worst of all the relation of membership, are profoundly mysterious. Mysteries are an onerous burden. Should we dump the burden by dumping the classes? If classes do not exist, we needn't puzzle over them. Renounce classes, and we are set free.

No; for set theory pervades modern mathematics. Some special branches and some special styles of mathematics can perhaps do without, but most of mathematics is into set theory up to its ears. If there are no classes, then there are no Dedekind cuts, there are no homeomorphisms, there are no complemented lattices, there are no probability distributions,. . . . For all these things are standardly defined as one or another sort of class. If there are no classes, then our mathematics textbooks are works of fiction, full of false 'theorems'. Renouncing classes means rejecting mathematics. That will not do. Mathematics is an established, going concern. Philosophy is as shaky as can be. To reject mathematics for philosophical reasons would be absurd. If we philosophers are sorely puzzled by the classes that constitute mathematical reality, that's our problem. We shouldn't expect mathematics to go away to make our life easier. Even if we reject mathematics gently – explaining how it can be a most useful fiction, 'good without being true' – we still reject it, and that's still absurd. Even if we hold onto some mutilated fragments of mathematics that can be reconstructed without classes, if we reject the bulk of mathematics that's still absurd.

That's not an argument, I know. But I laugh to think how *presumptuous* it would be to reject mathematics for philosophical reasons. How would *you* like to go and tell the mathematicians that they must change their ways, and abjure countless errors, now that *philosophy* has discovered that there are no classes? Will you tell them, with a straight face, to follow philosophical argument wherever it leads? If they challenge your credentials, will you boast of philosophy's other great discoveries: that motion is impossible, that a being than which no greater can be conceived cannot be conceived not to exist, that it is unthinkable that anything exists outside the mind, that time is unreal, that no theory has ever been made at all probable by evidence (but on the other hand that an empirically ideal theory can't possibly be false), that it is a wide-open scientific question whether anyone has ever believed anything, and so on, and on *ad nauseam?* Not me!

There is a way out of our dilemma. Or part-way out, and that may have to be good enough. We can take a 'structuralist' line about the theory of singleton functions; and derivatively, a structuralist line about set theory (like that of Paul Fitzgerald, [4]) and about all the mathematics that reduces to set theory. Even if we don't grasp *the* member-to-singleton function, we can still understand what it is to be *a* singleton function: a function that has the right formal character, and also obeys whatever 'unofficial' axioms we see fit to accept. So we might take set theory to be not the theory of *the* singleton function, plus mereology, but rather the general theory of all singleton functions. A set-theoretical truth would then have the covertly universal form: for any singleton function s, —s—s—. We need not say that any one singleton function has any special status.

Compare an algebraist's answer to a protesting student who says he hasn't been told what the Klein 4-group is just by being shown the table for it:

$$\begin{array}{c|cccc} & e & a & b & c \\ \hline e & e & a & b & c \\ a & a & e & c & b \\ b & b & c & e & a \\ c & c & b & a & e \end{array}$$

What *are* these four things e, a, b, c? The Prof may answer: 'They're anything you like. No one thing is *the* Klein 4-group; rather, any function (or equivalently, any four-things-and-a-function) that obeys the table is *a* Klein 4-group. Anything I tell you about "the" Klein 4-group is tacitly general. For instance, when I said that any permutation of the non-identity elements is an automorphism, I meant this to go for any Klein 4-group, no matter what its elements might be. And there do exist Klein 4-groups' – and here he offers an example or two – 'so there's no fear that generalisations about Klein 4-groups will come out vacuously true'.

Similarly, we might well be attracted to a 'structuralist' philosophy of arithmetic. It says that there's no one sequence that is *the* number sequence; rather, arithmetic is the general theory of omega-sequences. Each sequence has its own zero, its own successor function, and so on. The successor function and the sequence are interdefinable, so equivalently we could say that arithmetic is the general theory of successor functions. A successor function is characterised by means of the Peano axioms, as follows.

A *successor function* is any unary one-one function s such that

(1) the domain of s consists of its range and one other thing (called the *s-zero*); and
(2) all things in the domain of s are generated from the *s*-zero by iterated applications of s.

(Clause (2) means: if there are some things, and the *s*-zero is one of them, and when x is one of them so is $s(x)$, then everything in the domain of s is one of them.) Arithmetical truths are general; they apply to all successor functions. An arithmetical truth has the covertly universal form: for any successor function s, —s—s—. There do exist successor functions, as witness the familiar set-theoretical models of arithmetic: the Zermelo numbers, the von Neumann numbers, etc. So again there is no fear that our generalisations will turn out vacuous and make the wrong things come out as arithmetical truths.

(By 'structuralism' I don't mean to suggest that the subject matter of arithmetic is some interesting entity, an 'abstract structure', common to all the many successor functions. I suspect such entities are trouble, but in any case, they're an optional extra. We needn't believe in 'abstract structures' to have general structural truths about all successor functions.)

Similarly, *mutatis mutandis* for singleton functions. We can characterise them by means of a modified, mereologized form of the Peano axioms:

A *singleton function* is any unary one-one function s such that

(0) the range of s consists of atoms (called *s-singletons*);
(1) the domain of s consists of all small fusions of *s*-singletons together with all things (called *s-individuals*) that have no *s*-singletons as parts; and
(2) all things are generated from the *s*-individuals by iterated applications of s and of fusion.

(Something is *small* iff its atoms do not correspond one-one with all the atoms. Clause (2) means: if there are some things, and every *s*-individual is one of them, and when x is one of them so is $s(x)$, and every fusion of some of them is one of them, then everything is one of them.)

Suppose we have some 'unofficial axioms' (credible ones, not the overbold ones that say the sets lack whereabouts and character) that tell us which are the things – cats, puddles, spacetime points, souls if there are any – that ought to turn out to be individuals. Suppose also that these axioms endorse our Division, Priority, and Fusion Theses. Then we may call a singleton function *s correct* iff its demarcation of individuals – that is, its division between *s*-individuals and fusions of *s*-singletons – falls just where the unofficial axioms say it should. Since we have decided that the null set shall be the fusion of all individuals, we can also put it this way: the unofficial axioms determine what ought to be the null set, and a correct singleton function is one that defines the null set in agreement with that determination. Let us ignore all other singleton functions, and say that set theory consists of the general truths about all correct singleton functions.

Structuralist set theory is nominalistic set theory, in the special sense of Goodman ([5], II.3). It has no set-theoretical primitive, neither the general relation of membership nor the special case of membership in singletons. All the 'combining' or 'collecting' or 'gathering together' in set theory is purely mereological. But we do not renounce classes. We still have things that are classes relative to all the many correct singleton functions. Although these functions disagree about which classes have which members, at least they agree about which things are the classes and which are the individuals.

If we go structuralist about singleton functions, we may bid farewell to all our worries about how we understand 'the' singleton function. That which is not there need not be understood. Sad to say, we do *not* bid farewell to our lamentable ignorance of the whereabouts and character of the classes.

If we go structuralist, do we rebel against established set theory, and all of set-theoretical mathematics? Well, we challenge no proofs and we deny no theorems. But we do rebuke the mathematicians for a foundational error: they think they've fixed on one particular singleton function when really all they've got is the general notion of *a* singleton function. Even this much of a philosophical challenge to established mathematics is presumptuous and suspect. It would be much better if we could find a way to take mathematics just as we find it. But I have no way to offer, and the structuralist reinterpretation is the best fallback I know.

You may smell a rat. Here we are, quantifying over functions. How can we do that before we already have set theory, where 'having' set theory includes understanding its primitive notions? Isn't a function a class of ordered pairs? And aren't ordered pairs, in turn, set-theoretical constructions: Kuratowski pairs, or something similar? And when I characterised singleton functions, didn't I quantify over relations again when I spoke of one-one correspondence? And wasn't it set-theoretic modelling that gave us our needed assurance that there did exist Klein 4-groups and successor functions, and don't we need a parallel assurance that there exist singleton functions? In short, structuralism about set theory seems to presuppose set theory. Only if we don't need it can we have it.

Recent work by John P. Burgess and A. P. Hazen saves the day. They've shown how we can simulate quantification over relations using only the framework of megethology: mereology plus plural quantification. Roughly speaking, a quantifier over relations is a plural quantifier over things that encode ordered pairs by mereological means.

Here we shall take a hybrid method, beginning Hazen's way and finishing Burgess's way. We must assume that there are infinitely many atoms; at this stage of the game, we can express that by saying that there are some things, each of which is part of another. For simplicity, we shall also assume that Reality consists entirely of atoms. (But if it did not, indeed even if there were no atoms at all, we could assume instead just that there are infinitely many mutually non-overlapping things – call them 'quasi-atoms' – whose fusion is all of Reality. Then all that follows could be unchanged, except that 'atom' would mean 'quasi-atom' and all quantifications would be tacitly restricted to things that were fusions of quasi-atoms.)

First step (Zermelo). We can encode orderings of atoms. Given some things O, atom b *O-precedes* atom c iff whenever c is part of one of O then b is too, but not conversely. Say that O *order the atoms* iff O-precedence is transitive, asymmetric, and connected. We assume as a principle of megethology that some O order the atoms.

Second step (Hazen). We can encode relations of atoms. We can say that O, A, B, C *relate atoms* iff O order the atoms, A are some diatoms (two-atom fusions), B are some diatoms, and C are some atoms. Given some such O, A, B, C, say that atom b is *OABC-related* to atom c iff either b O-precedes c and $b + c$ is one of A, or c O-precedes b and $b + c$ is one of B, or b and c are identical and b is one of C. The idea is that the extraneous ordering O takes the place of the ordering normally built into ordered pairs, so that we can get by with *un*ordered diatoms. (We would need to make special provision for some relations: in the case of the identity relation, for instance, both A and B would be missing. But we can encode the two relations needed for the next step, and that is enough.) Consider a special case; O, A, B, C *map all atoms one-one* iff O, A, B, C relate atoms; any atom is $OABC$-related to exactly one atom, called its *OABC-image*; and no two atoms have the same $OABC$-image. We define the *OABC-image* of anything as the fusion of $OABC$-images of its atoms.

Third step (Burgess). We can encode ordered pairs. We assume as a principle of megethology that if there are infinitely many atoms, then there are O, A, B, C, D, E, F such that O, A, B, C map all atoms one-one, and so do O, D, E, F; and such that no $OABC$-image ever overlaps any $ODEF$-image. Thus all of Reality is mapped into two non-overlapping microcosms, and every part of Reality has an image in each microcosm. The *ordered pair* of x and y (with respect to $O \ldots F$) is the fusion of the $OABC$-image of x and the $ODEF$-image of y. We recover the first term of an ordered pair z as the fusion of all atoms whose $OABC$-images are parts of z; and the second term as the fusion of all atoms whose $ODEF$-images are parts of z.

Final step. We simulate a quantifier over relations (binary relations will suffice for present purposes) by a plural quantifier over ordered pairs, preceded by a string of plural quantifiers over the wherewithal for decoding such pairs. Example. 'For some relation r, ——$r(x, y)$ ——' becomes: 'For some $O \ldots F$ meeting the conditions above, for some R such that each one of R is an ordered pair (with respect of O ... F), ——the ordered pair (with respect to O ... F) of x and y is one of R——.' The case of a universal quantifier over relations is similar, except that all the quantifiers in the string are universal.

Megethology therefore suffices as a framework for structuralist set theory. Once we have simulated quantification over relations, we can define singleton functions; we can quantify over them; and with respect to any given one of them, we can define membership, classhood, etc. as above.

Now we shall formulate some hypotheses about the size of Reality, and thereby see how megethology deserves its name. Recall that something is *small* iff its atoms correspond one-one with some but not all the atoms; otherwise *large*. Similarly in the plural: some things are *few* iff they correspond one-one with some but not all the atoms, otherwise *many;* and they are *barely many* iff they correspond one-one with all the atoms. An *infinite* thing is one whose atoms correspond one-one with only some of its atoms.

Hypothesis U. *The fusion of a few small things is small.*

Hypothesis P—. *The parts of a small thing are few or barely many.*

Hypothesis P. *The parts of a small thing are few.*

Hypothesis I. *Something small is infinite.*

Hypotheses U and P— together imply this

Existence Thesis. *For any small thing, n, there is a singleton function s such that n is the fusion of the s-individuals.*

First Step. Given Hypotheses U and P—, we have that the small parts of Reality are barely many. Proof of the first step: Assume, as a principle of megethology, that we have a well-ordering of all the atoms. If so, then we have an *initial* well-ordering of all the atoms: that is, one in which each atom is preceded by only a few others. For, given a well-ordering that isn't already initial, we take the first atom that is preceded by many others. Imaging under the one-one correspondence between these many preceding atoms and all the atoms, we have a new well-ordering of all the atoms; and this new well-ordering, being isomorphic to a segment of the old one in which each atom is preceded by only a few others, is initial. Atom b *bounds* x iff every atom of x precedes b in our initial well-ordering. Any small x is bounded: for any atom c of x, let $[c]$ be the fusion of c and all atoms that precede c; these $[c]$'s are small and few; their fusion is small, by Hypothesis U; some atom falls outside that fusion, and it bounds x. If any atom bounds x, there is a first atom that bounds x, called the *limit* of x. Any small x has a limit. When c is the limit of x, x is part of $[c]$. By P $-$, since $[c]$ is small, $[c]$'s parts are at most barely many. So each of barely many atoms is the limit of at most barely many small things. So, by a principle of megethology akin to the multiplication rule for cardinal numbers, there are at most barely many small things. And each of barely many atoms is small, so there are at least barely many small things.

Second Step. Given that the small parts of Reality are barely many, we have that for any small thing n, there exists a unary one-one function f such that (0) the range of f consists of atoms of $-n$; (1) the domain of f consists of all small parts of $-n$ together with all parts of n. Proof of the second step: By a principle of megethology, all parts of n are themselves small. So the small parts of $-n$ together with the parts of n are some of the small things, so they are at most barely many. And they are at least barely many, since all atoms are either small parts of $-n$ or parts of n. By another principle of megethology and the infinity of atoms, since n is small, $-n$ is large. So the domain and range of the desired function both stand in one-one correspondence with all the atoms, so they stand in one-one correspondence with each other.

Third Step (due partly to Burgess). Given f as specified in the second step, there is a singleton function s such that the s-individuals are exactly the parts of n. Proof of the third step: Though f needn't be a singleton function, call $f(y)$ the *-singleton* of y; define *-classes, *-sets, and *-membership accordingly. Let G be the things generated from the parts of n by iterated applications of f and of fusion. Let H be the *-singletons among G. Now consider the *-class of all *-sets among G that are non-self-*-members. It cannot itself be a *-set among G, by Russell's paradox. But if it were small, it would be a *-set among G. So it is large. So some of H are many. We saw that $-n$ is large; so H stand in one-one correspondence with all the atoms of $-n$. Extend this to a correspondence that also maps all atoms of n to themselves. The image of f under this correspondence is the desired function s. QED

That is, there is a correct singleton function, no matter what small thing the unofficial axioms may deem to be the null set. And we have this

Categoricity Thesis. *For any small thing n, if we have two single-ton functions s and t such that n is the fusion of the s-individuals and also the fusion of the t-individuals, then s and t differ only by a permutation of singletons.*

Proof. Define relation p as follows: if x is an individual (part of n), $s(x)$ bears p to $t(x)$; if x and y are small fusions of singletons (atoms of $-n$), and every atom of x bears p to some atom of y, and for every atom of y, some atom of x bears p to it, then $s(x)$ bears p to $t(y)$. Call a single-ton *good*-1 iff it bears p to exactly one singleton; and call anything *good*-1 iff every singleton that is part of is good-1. Likewise, call a singleton *good*-2 iff exactly one singleton bears p to it; and call any-thing *good*-2 iff every singleton that is part of it is good-2. Using clause (2) in the definition of a singleton function, we can show by induction that everything is good-1, and that everything is good-2. So p is a permutation of singletons; and t is the image of s under p. QED

Take any mathematical – that is, set-theoretical – sentence. Its only vocabulary, after we eliminate defined terms, will be logical and mereological vocabulary, and 'singleton'. Replace 'singleton' by a variable to obtain a formula —s—s—. The original sentence is math-ematically true iff this formula holds for all correct singleton func-tions; mathematically false iff it holds for none; and indeterminate iff it holds for some but not others. The Existence Thesis guarantees that nothing is both mathematically true and mathematically false. The Categoricity Thesis guarantees that there will be no indetermi-nacy; for if two singleton functions differ only by a permutation, then both or neither will satisfy the formula. Despite our structuralist

226

reinterpretation, mathematical sentences come out bivalent, just as if we had been able to fix on one particular singleton function.

Given our definitions and some principles of megethology, we can show that all but two of the standard axioms of set theory hold for any singleton function: Null Set, Extensionality, Pair Sets, *Aussonderung*, Replacement, *Fundierung*, Choice, and Unions. We get the last two standard axioms only conditionally: Power Sets given Hypothesis P, and Infinity given Hypothesis I.

Proof. Null Set holds because, by definition, there are individuals; their fusion exists, by the principle of Unrestricted Composition; and by our definitions it is a set with no members.

Extensionality holds in virtue of a principle of Uniqueness of Composition: it never happens that the same things have two different fusions. Apply it to classes: there are never two fusions of the same singletons, hence no coextensive classes. Apply it to individuals: there are not two fusions of the individuals, hence there is only one null set.

Pair Sets follows from Unrestricted Composition, plus the fact that there are at least three atoms (singleton of the null set, singleton of that, singleton of that) and hence any fusion of two singletons is small.

Aussonderung follows from Unrestricted Composition plus the fact that any part of a small thing is small (the Null Set axiom covers the case in which the required set is empty).

Replacement follows from a corresponding principle of megethology: if there is a function from atoms of x to all atoms of y, then y is small if x is.

Fundierung. Something is *grounded* iff it belongs to no class that intersects each of its own members and thereby violates *Fundierung*. Using clause (2) of the definition of a singleton function, we can show by induction that everything is grounded, hence nothing violates *Fundierung*.

Choice follows from a corresponding principle of megethology: given some non-overlapping things, something shares exactly one atom with each of them.

Unions. First we need a Lemma: the existence of a singleton function implies Hypothesis U. Proof of the Lemma: If not, we have a few small things *R* such that their fusion is not small. Then the atoms of

227

the fusion of R correspond one-one with all the atoms. Let the things S be the images of R under this correspondence; then S are a few small things, but their fusion is all of Reality. Given a singleton function, let the things T be as follows: whenever a class or individual is one of S, it is one of T; whenever a mixed fusion is one of S, its largest individual part is one of T and its largest class part is one of T. T also are a few small things, their fusion is all of Reality, and further each one of them has a singleton. Now we can encode anything x, unambiguously, by an atom: first we have the intersections of x with each of T, and these are a few individuals and sets; then we have the set of these intersections; and finally we have the singleton of that set. So there are no more fusions of atoms than atoms, which is impossible by the reasoning of Cantor's theorem.

Now, take any set. If it has no classes as members, its union is the null set. Otherwise we have its union class, by Unrestricted Composition; the Lemma gives us Hypothesis U, whereby this class is small and hence a set.

Power Sets. For any set, we have its power class, by Unrestricted Composition. Given Hypothesis P, this class is small, hence a set.

Infinity. Given Hypothesis I, something infinite is small. We have the class of its atoms, by Unrestricted Composition and the fact that all atoms are either individuals or singletons, and this class is an infinite set. QED

So to regain set theory we need to assume Hypotheses U, P, and I. (We needn't list P– separately, since it follows from P.) These constrain the size of Reality, as measured by the total number of atoms. It is easy to see how any two of the constraints can hold. U and P hold, but I fails, if there are countably many atoms, so that 'small' means 'finite'. U and I hold, but P fails, if there are aleph-one atoms, so that 'small' means 'countable'. P and I hold, but U fails, if there are beth-omega atoms. Making all three hold together is harder. That takes a (strongly) 'inaccessible' infinity of atoms – an infinity that transcends our commonplace alephs and beths in much the same way that any infinity transcends finitude. There will be inaccessibly many atoms, inaccessibly many singletons, and inaccessibly many sets.

Do you find it extravagant to posit so many things? Especially when you know nothing about their whereabouts and character? Beware! The inaccessibly many atoms are not the wages of a newfangled mereological-cum-structuralist reconstruction of set theory. It is orthodox set theory itself that incurs the commitment to an inaccessible domain of sets. Like this: (1) The largest classes are proper classes, smaller ones are sets. (2) The size of the proper classes cannot be reached from below by taking unions: the union of a set of sets is still a mere set. (3) The size of the proper classes cannot be reached from below by taking powers: the class of all subsets of a set is still a mere set. (4) The size of the proper classes is not the smallest infinite size: some mere set is infinite. So says orthodoxy. I have faithfully reconstructed this aspect of orthodoxy along with the rest. Will you tell the mathematicians to abjure their errors? Not me!

REFERENCES

1 George Boolos, 'To be is to be the value of a variable (or to be some values of some variables)', *Journal of Philosophy*, **81** (1984), 430–49.
2 ———, 'Nominalist Platonism', *Philosophical Review*, **94,** (1985), 327–44.
3 John P. Burgess, A. P. Hazen, David Lewis, 'Appendix on pairing' in [6].
4 Paul Fitzgerald, 'Meaning in science and mathematics', *PSA 1974*, R. S. Cohen *et al.,* eds. (Reidel, 1976), section IV.
5 Nelson Goodman, *The structure of appearance* (Harvard University Press, 1951).
6 David Lewis, *Parts of classes* (Blackwell, 1991).

Index

232

Kalman implication, 99
Kamp, Hans, 43, 51, 56
Kaplan, David, 21n, 36, 40–44, 180n
Klein 4-groups, 219, 222
Kratzer, Angelika, 77–96
Kripke, Saul, 42

Loewer, Barry, 86, 96
Lucas, J.R., 3, 166–173

Magpie, 155, 204, 207, 209, 212
Makinson, D.C., 98, 99, 109
mathematics, rejection of, 217–218,
 221–222, 229
mechanism, 3, 166–173
megethology, 4, 203–229
mereology, 4, 180–185, 186–202,
 203–229
 of subject matters, 112–113,
 116–117, 140, 146–149
Meyer, Robert, 67n, 98, 109, 122,
 124
modal logics, 2, 66, 75
Montague, Richard, 23, 44

Nolan, Daniel, 208
nominalism, Goodmanian, 186–202,
 221; see also rejection of
 mathematics
non-iterative axioms, 66–76
null set, 206, 207, 210–211
numerative sentences, 181–185

observation, 125–158
ordering frames, 78–80, 84–95

paraconsistency, 2–3, 54, 97–110
Parsons, Terence, 25, 44
Peano axioms, 167–169, 171–172
philosophy, follies of, 218
Pinter, Charles, 106, 109
Plumwood, Valerie, 98, 110, 122,
 124

plural quantification, 203, 213–214,
 222–224
Pollock, John, 77–96
positivism, logical, 3, 125–158
Possum, 155, 204, 207, 209, 211, 212
 216
premise frames, 81–85, 87, 89–91,
 93–94
Priest, Graham, 98, 100–101, 107,
 110, 111n, 121, 122, 124
probability, 2, 57–65, 150–153
proper classes, 206, 212–213, 229
pseudo-sets, 186–202

quality classes, 176–179
quantification, adverbs of, 2, 5–19
 plural, 203, 213–214, 222–224
 simulated, 203, 204, 222–224
 variables of, 8–19, 33
questions, 46–48, 112, 114, 120
Quine, W.V., 33, 44, 190, 202
quodlibet, 98–100, 104, 105, 120–123,
 133n

Railton, Peter, 139n
rejection of mathematics, 217–218,
 221–222, 229
relevance, 2–3, 97–110, 111–124
 to observation, 125–155
Rescher, Nicholas, 103, 110
Routley, Richard, 67n, 98, 100–101,
 102, 110, 122, 124
Routley, Valerie, 98, 110, 122,
 124
Russell, Bertrand, 10n, 20, 101
 Russell's paradox, 101, 212–214,
 225

Scheffler, Israel, 158
Schotch, Peter, 103, 110
science, 125–155, 162, 174
Searle, John, 48
Segerberg, Krister, 51, 56